Nitric Oxide in Plant Physiology

Edited by
Shamsul Hayat, Masaki Mori,
John Pichtel, and Aqil Ahmad

Related Titles

Hirt, H. (ed.)

Plant Stress Biology

From Genomics to Systems Biology

2010

ISBN: 978-3-527-32290-9

Kahl, G., Meksem, K. (eds.)

The Handbook of Plant Functional Genomics

Concepts and Protocols

2008

ISBN: 978-3-527-31885-8

Ahmad, A., Pichtel, J., Hayat, S. (eds.)

Plant-Bacteria Interactions

Strategies and Techniques to Promote Plant Growth

2008

ISBN: 978-3-527-31901-5

Scott, P.

Physiology and Behaviour of Plants

2008

ISBN: 978-0-470-85025-1

Buchanan, B., Gruissem, W., Jones, R. L. (eds.)

Biochemistry & Molecular Biology of Plants

2002

ISBN: 978-0-943088-39-6

Nitric Oxide in Plant Physiology

Edited by
Shamsul Hayat, Masaki Mori, John Pichtel,
and Aqil Ahmad

WILEY-
BLACKWELL

WILEY-VCH Verlag GmbH & Co. KGaA

The Editors

Dr. Shamsul Hayat
Aligarh Muslim University
Department of Botany
Aligarh 202002
India

Dr. Masaki Mori
National Institute of Agrobiological Sciences
Division of Plant Sciences
2-1-2 Kannondai, Tskuba
Ibaraki 305-8602
Japan

Prof. John Pichtel
Ball State University
Department of Natural Resources and
Environmental Management
WQ 103
Muncie, IN 47306
USA

Dr. Aqil Ahmad
Higher College of Technology
Department of Applied Sciences
Al-Khuwair
Sultanate of Oman
Oman
and
Aligarh Muslim University
Department of Botany
Aligarh 202002
India

Library of Congress Card No.: applied for

British Library Cataloguing-in-Publication Data
A catalogue record for this book is available from the British Library.

Bibliographic information published by the Deutsche Nationalbibliothek
The Deutsche Nationalbibliothek lists this publication in the Deutsche Nationalbibliografie; detailed bibliographic data are available on the Internet at http://dnb.d-nb.de.

Cover Formgeber, Eppelheim
Typesetting Thomson Digital, Noida, India
Printing and Binding Strauss Gmbh, Mörlenbach

Printed in the Federal Republic of Germany
Printed on acid-free paper

ISBN: 978-3-527-32519-1

Contents

Nitric Oxide in Plant Physiology. Edited by S. Hayat, M. Mori, J. Pichtel, and A. Ahmad

ISBN: 978-3-527-32519-1

Preface

The role of nitric oxide (NO) in biological systems has experienced increased prominence in the scientific literature since the 1980s, particularly after coming to light as a signaling molecule in plants in the late 1990s. The number of publications concerning the influence of nitric oxide in plants has dramatically increased since then. Nitric oxide is an easily diffused bioactive and signal-transmitting molecule that directly regulates many plant functions including germination, leaf expansion, root growth, stress physiology, and sequential cell death. This molecule also participates in the adaptation of plants to environmental stresses, working as the key signal carrier in defense response. Recent studies have shown that nitric oxide imparts synergistic effects with phytohormones in physiological regulation and signal transmission.

The purpose of this work is to present recent advances in the role on nitric oxide on plant physiology. This book, composed of 13 chapters contributed by scholars worldwide, addresses mechanisms of NO action in specific plant physiological processes and application of instrumentation for assessing such actions. Chapter 1 embodies recent discoveries in NO chemical properties and the mechanism of NO biosynthesis with special emphasis on the role of nitric oxide in physiological and biochemical changes that occur in plants under normal conditions due to exogenously applied or endogenously produced nitric oxide. Also presented is the issue of cross talk between nitric oxide and other phytohormones. In Chapter 2, electron paramagnetic resonance (EPR) is discussed as a tool to study nitric oxide generation in plants. Recent progress in nitric oxide research with respect to calcium and signaling is discussed in Chapter 3. The current knowledge of nitric oxide in plants exposed to diverse environmental stresses such as salinity, heavy metals, UV-B radiation, ozone, and mechanical wounding comprises Chapter 4. Chapter 5 deals with nitric oxide biosynthesis in relation to polyamines and cytokinins. Moreover, there are a number of similarities in cytokinin- and polyamine-mediated physiological functions such as embryogenesis, flowering, and senescence and in plant responses to abiotic and biotic stresses, which indicate overlapping functions of both these signaling substances; this is also discussed in Chapter 5.The role of nitric oxide in cadmium-induced PCD is discussed in Chapter 6, suggesting a possible

Nitric Oxide in Plant Physiology. Edited by S. Hayat, M. Mori, J. Pichtel, and A. Ahmad

ISBN: 978-3-527-32519-1

regulatory role in response to heavy metal stress. Chapter 7 focuses on how research on nitrate reductase-deficient plants may contribute to the elucidation of mechanisms involved in nitric oxide production and signaling during plant–pathogen interactions. The function of nitric oxide in plant–pathogen interactions is discussed in detail in Chapter 8. In Chapter 9, various nitric oxide functions as a bioactive signaling molecule in plant abiotic stress responses are discussed. The cross talk between NO and other key signaling components under abiotic stress is also reviewed. The role of NO in the recovery from disease related to the stimulation of wound-healing processes is discussed in Chapter 10. Nitric oxide signaling by plant-associated bacteria is discussed in Chapter 11. Chapter 12 deals with nitric oxide synthase (NOS) activity in pea. The possibility of NO production from various sources in pea cells is also discussed. Chapter 13 describes posttranslational modifications of protein by nitric oxide.

This book is not intended to serve as an encyclopedic review of the subject of NO and its role in plant physiology; however, the various chapters incorporate both theoretical and practical aspects and serve as a baseline information for future research. It is hoped that this book will be useful to students, teachers, and researchers, both in universities and in research institutes, especially in the field of biological and agricultural sciences.

With great pleasure, we extend our sincere thanks to all contributors for their timely response, their excellent and up-to-date contributions, and consistent support and cooperation. We are also thankful to Dr. Zaki A. Siddiqui and Dr. Qazi Fariduddin, Department of Botany, Aligarh Muslim University, Aligarh, India for their encouragement. We are extremely thankful to Wiley-VCH, Germany, for expeditious acceptance of our proposal and completion of the review process. Subsequent cooperation from Wiley-VCH staff is also gratefully acknowledged. We express our sincere thanks to the members of our families for all the support they provided and the neglect and loss they suffered during the preparation of this book.

Finally, we are thankful to the Almighty who provided and guided all the channels to work in cohesion and coordination right from the conception of the idea to the development of the final version of this treatise, "Nitric Oxide in Plant Physiology."

Shamsul Hayat
Masaki Mori
John Pichtel
Aqil Ahmad

List of Contributors

Jasmeet Kaur Abat
University of Delhi
Department of Botany
Plant Molecular Physiology and
Biochemistry Laboratory
Delhi 110007
India

Aqil Ahmad
Higher College of Technology
Department of Applied Sciences
Al-Khuwair
Sultanate of Oman

and

Aligarh Muslim University
Department of Botany
Aligarh 202002
India

Magdalena Arasimowicz-Jelonek
Adam Mickiewicz University
Faculty of Biology
Department of Plant Ecophysiology
Umultowska 89
61-614 Poznan
Poland

Juan B. Barroso
Universidad de Jaén
Área de Bioquímica y Biología
Molccular
Unidad Asociada al CSIC (EEZ)
Grupo de Señalización Molecular y
Sistemas Antioxidantes en Plantas
Jaén
Spain

Alejandro D. Boveris
University of Buenos Aires
School of Pharmacy and Biochemistry
Physical Chemistry-PRALIB
Buenos Aires
Argentina

Marcia Regina Braga
Instituto de Botânica (IBt)
Seção de Fisiologia e Bioquímica de
Plantas
CP 3005
01061-970 São Paulo, SP
Brazil

Michael F. Cohen
Sonoma State University
Department of Biology
Rohnert Park, CA
USA

Nitric Oxide in Plant Physiology. Edited by S. Hayat, M. Mori, J. Pichtel, and A. Ahmad

ISBN: 978-3-527-32519-1

Francisco J. Corpas
Estación Experimental del Zaidín (EEZ), CSIC
Departamento de Bioquímica, Biología Celular y Molecular de Plantas
Granada
Spain

Alex Costa
University of Padova
Department of Biology
Via U. Bassi 58/B
35131 Padova
Italy

Eric L. Davis
North Carolina State University
Department of Plant Pathology
Raleigh, NC 27695-7616
USA

Luis A. del Río
Estación Experimental del Zaidín (EEZ), CSIC
Departamento de Bioquímica, Biología Celular y Molecular de Plantas
Granada
Spain

Roberto De Michele
University of Padova
Department of Biology
Via U. Bassi 58/B
35131 Padova
Italy

Halley Caixeta de Oliveira
Universidade Estadual de Campinas – UNICAMP
Instituto de Biologia
Departamento de Bioquímica
CP 6109
13083-970 Campinas, SP
Brazil

Renu Deswal
University of Delhi
Department of Botany
Plant Molecular Physiology and Biochemistry Laboratory
Delhi 110007
India

Qazi Fariduddin
Aligarh Muslim University
Department of Botany
Plant Physiology Section
Aligarh 202002
India

José A. Feijó
Instituto Gulbenkian de Ciência
Centro de Biologia do Desenvolvimento
Oeiras
Portugal

and

Universidade de Lisboa
Faculdade de Ciências
Dept Biologia Vegetal
Campo Grande
Lisboa
Portugal

Jolanta Floryszak-Wieczorek
Poznan University of Life Sciences
Department of Plant Physiology
Wołyńska 35
60-637 Poznan
Poland

Gang-Ping Hao
Taishan Medical University
Department of Biological Science
Taian 271000
China

Syed Aiman Hasan
Aligarh Muslim University
Department of Botany
Plant Physiology Section
Aligarh 202002
India

Shamsul Hayat
Aligarh Muslim University
Department of Botany
Plant Physiology Section
Aligarh 202002
India

Jengsheng Huang
North Carolina State University
Department of Plant Pathology
Raleigh, NC 27695-7616
USA

Sebastián Jasid
University of Buenos Aires
School of Pharmacy and Biochemistry
Physical Chemistry-PRALIB
Buenos Aires
Argentina

Lorenzo Lamattina
Universidad Nacional de Mar del Plata
Instituto de Investigaciones Biológicas
Facultad de Ciencias Exactas y Naturales
CC 1245
7600 Mar del Plata
Argentina

Marina Leterrier
Estación Experimental del Zaidín (EEZ), CSIC
Departamento de Bioquímica, Biología Celular y Molecular de Plantas
Granada
Spain

Fiorella Lo Schiavo
University of Padova
Department of Biology
Via U. Bassi 58/B
35131 Padova
Italy

Luzia V. Modolo
Universidade Federal de Minas Gerais
Instituto de Ciências Biológicas
Departamento de Botânica
Belo Horizonte, MG 31270-901
Brazil

Masaki Mori
National Institute of Agrobiological Sciences
Division of Plant Sciences
Plant Disease Resistance Unit
2-1-2, Kannondai, Tsukuba
Ibaraki 305-8602
Japan

José M. Palma
Estación Experimental del Zaidín (EEZ), CSIC
Departamento de Bioquímica, Biología Celular y Molecular de Plantas
Granada
Spain

D. Marshall Porterfield
Purdue University
Bindley Bioscience Center
Physiological Sensing Facility
Discovery Park
West Lafayette, IN
USA

and

Purdue University
Department of Agricultural & Biological Engineering
West Lafayette, IN
USA

and

Purdue University
Department of Horticulture & Landscape Architecture
West Lafayette, IN
USA

and

Purdue University
Weldon School of Biomedical Engineering
West Lafayette, IN
USA

Ana Margarida Prado
Instituto Gulbenkian de Ciência
Centro de Biologia do Desenvolvimento
Oeiras
Portugal

Susana Puntarulo
University of Buenos Aires
School of Pharmacy and Biochemistry
Physical Chemistry-PRALIB
Buenos Aires
Argentina

Ione Salgado
Universidade Estadual de Campinas – UNICAMP
Instituto de Biologia
Departamento de Bioquímica
CP 6109
13083-970 Campinas, SP
Brazil

Günther F.E. Scherer
Leibniz Universität Hannover
Institut für Zierpflanzenbau und Gehölzforschung
Abt. Molekulare Ertragsphysiologie
Herrenhäuser Str. 2
30419 Hannover
Germany

Marcela Simontacchi
University of Buenos Aires
School of Pharmacy and Biochemistry
Physical Chemistry-PRALIB
Buenos Aires
Argentina

Serenella Sukno
Universidad de Salamanca
Centro Hispano-Luso de Investigaciones Agrarias (CIALE)
Departamento de Microbiología y Genética
Campus de Villamayor C/Río Duero 12
37185 Villamayor
España

Yee-How Tan
Universiti Putra Malaysia
Faculty of Agriculture
Department of Plant Protection
43400 Serdang
Malaysia

Rinukshi Wimalasekera
Leibniz Universität Hannover
Institut für Zierpflanzenbau und Gehölzforschung
Abt. Molekulare Ertragsphysiologie
Herrenhäuser Str. 2
30419 Hannover
Germany

Mui-Yun Wong
Universiti Putra Malaysia
Institute of Tropical Agriculture
43400 Serdang
Malaysia

and

Universiti Putra Malaysia
Faculty of Agriculture
Department of Plant Protection
43400 Serdang
Malaysia

Hideo Yamasaki
University of the Ryukyus
Faculty of Science
Nishihara, Okinawa 903-0213
Japan

Jianhua Zhang
Hong Kong Baptist University
Department of Biology
Hong Kong
China

Michela Zottini
University of Padova
Department of Biology
Via U. Bassi 58/B
35131 Padova
Italy

and

Universidad Autònoma de Barcelona
Departamento de Bioqumica y Biologa Molecular
08193 Bellaterra, Barcelona
Spain

1
Nitric Oxide: Chemistry, Biosynthesis, and Physiological Role

Shamsul Hayat, Syed Aiman Hasan, Masaki Mori, Qazi Fariduddin, and Aqil Ahmad

Summary

NO is recognized as a biological messenger in plants. It is a highly reactive gaseous free radical, soluble in water and lipid. It can be synthesized in plants via different enzymatic and nonenzymatic sources such as NOS, NR, XOR, and Ni-NOR. Due to its high lipophilic nature, it can easily diffuse through membrane and can act as a inter- and intracellular messenger and regulate diverse physiological and biochemical processes in plants in a concentration-dependent manner, such as seed dormancy, growth and development, senescence, respiration, photosynthesis, programmed cell death, antioxidant defense system, and so on. Moreover, NO also has an ability to act simultaneously with other molecules and signals in plants. This chapter covers the advances in chemical properties and mechanism of its biosynthesis with special emphasis on the role of NO in the physiological and biochemical changes that occur in plants under normal conditions due to the exogenously applied or endogenously produced NO, along with the cross talk between NO and other phytohormones.

1.1
Introduction

Since the past decade, nitric oxide (NO) is recognized as a novel biological messenger in plants and animals and has received special attention from most of the branches of biological sciences, including medicine, biochemistry, physiology, and genetics. The interest of biologists gained special momentum when this highly reactive radical was identified as a potent endogenous vasodilator of the endothelium [1]. Moreover, a widespread biological significance of nitric oxide was first recognized by Koshland [2] who named this free radical as "Molecule of Year." The 1998 Nobel Prize in Physiology and Medicine was awarded for the discovery of NO as a biological mediator produced in mammalian cells.

Nitric Oxide in Plant Physiology. Edited by S. Hayat, M. Mori, J. Pichtel, and A. Ahmad

ISBN: 978-3-527-32519-1

The role of NO is not confined only to the animal kingdom, but plants also have the ability to accumulate and metabolize atmospheric NO. Klepper [3] for the first time observed the production of NO in soybean plant, treated with photosynthetic inhibitor herbicides [4, 5], other chemicals [6, 7], or under anaerobic conditions [6, 8]. In plants, NO can be generated via enzymatic and nonenzymatic pathways. The enzymatic pathway is catalyzed by cytosolic nitrate reductase (cNR), NO synthase (NOS) or NOS-like enzymes, and nitrite:NO reductase (Ni-NOR). Nonenzymatic pathway is nitrite dismutation to NO and nitrate at acidic pH values [9–11].

After the discovery of the existence of NO in plant, the question arose, should NO be considered a phytohormone or not because the classical concept of hormone is based on three premises [12]: (i) localized site of biosynthesis, (ii) transport to target cells especially separated from the place of synthesis, and (iii) control of responses through changes in endogenous levels of the chemical. First, NO had been found to be formed mainly in actively growing tissues such as embryonic axes and cotyledons, and the level decreases in mature and senescing organs [13, 14]. Second, the smaller size of the molecule and its higher diffusion rate through biological membranes mean that NO fits the premise that hormones are easily transported. Third, it is the sensitivity of the target cells, rather than the concentration of the plant hormone, that defines the magnitude of a response [15]; because of this concept, some scientists decided to substitute the term hormone with a wider term, "plant growth regulator." Later on, Beligni and Lamattina [16] categorized NO as a nontraditional regulator of plant growth.

Further investigations led to the finding that NO is soluble in water and lipid. It can exist in three interchangeable forms: the radical ($NO^{\bullet}$), nitrosonium cation (NO^{+}), and nitroxyl anion (NO^{-}). Due to its high lipophilic nature, NO may diffuse through membranes [17] and acts as an inter- and intracellular messenger in many physiological functions. It plays a significant role in plant growth and development, seed germination, flowering, ripening, and senescence of organs [18]. Moreover, like other phytohormones, NO also acts in a concentration-dependent manner.

Research on NO in plants has gained a considerable attention in recent years and there is increasing evidence corroborating the role of this molecule in plants. Therefore, in this chapter, an effort has been made to cover the recent advances in chemical properties and mechanism of its biosynthesis with special emphasis on the role of NO in physiological and biochemical changes that occur in plants under normal conditions due to the exogenously applied or endogenously produced NO, along with the cross talk between NO and other phytohormones.

1.2
Nitric Oxide Chemistry

Nitric oxide is a gaseous free radical; its chemistry implicates an interplay between the three redox-related species: nitric oxide radical ($NO^{\bullet}$), nitrosonium cation (NO^{+}), and nitroxyl anion (NO^{-}). In biological systems, $NO^{\bullet}$ reacts rapidly with atmospheric oxygen (O_2), superoxide anion ($O_2^{\bullet -}$), and transition metals. The reaction of $NO^{\bullet}$ with O_2 results in the generation of NO_x compounds (including $NO_2^{\bullet}$, N_2O_3, and N_2O_4), which can either react with cellular amines and thiols or simply hydrolyze to form the end metabolites nitrite (NO_2^-) and nitrate (NO_3^-) [19]. The reaction of $NO^{\bullet}$ with $O_2^{\bullet -}$ yields peroxynitrite ($ONOO^-$), a powerful oxidant that mediates cellular injury. At physiological pH, $ONOO^-$ equilibrates rapidly with pernitrous acid (ONOOH) that, depending on its conformation, rapidly decomposes to NO_3^- or to the highly reactive hydroxyl radical $HO^{\bullet}$. $NO^{\bullet}$ also forms complexes with transition metals found in heme- or cluster-containing proteins, thus forming iron–nitrosyl complexes. This process alters the structure and function of the target proteins, as exemplified by the activation of soluble guanylate cyclase and the inhibition of aconitases.

In addition, $NO^{\bullet}$ is extremely susceptible to both oxidation and reduction. One-electron oxidation of $NO^{\bullet}$ leads to the formation of NO^+ (nitrosonium cation), while the product of one-electron reduction of $NO^{\bullet}$ is a nitroxyl anion (NO^-) [20–22]. This oxidation can be supported by Fe(III)-containing metalloproteins [20, 21]. NO^+ mediates electrophilic attack on reactive sulfur, oxygen, nitrogen, and aromatic carbon centers, with thiols being the most reactive groups. This chemical process is referred to as nitrosation. Nitrosation of sulfhydryl (*S*-nitrosation) centers of many enzymes or proteins has been described and the resulting chemical modification affects the activity in many cases. Such modifications are reversible and protein *S*-nitrosation–denitrosation could represent an important mechanism for regulating signal transduction. One-electron reduction of $NO^{\bullet}$ generates NO^-. The physiological significance of NO^- has not been clarified. Some workers [20, 23] suggest that it could act as a stabilized form of NO. NO^- is also believed to react with Fe(III) heme and to mediate sulfhydryl oxidation of target proteins.

1.3
Biosynthesis of Nitric Oxide

In biological systems, NO can be formed both enzymatically and nonenzymatically. The enzyme responsible for NO generation in animals is nitric oxide synthase. Although NOS-like activity has been detected widely in plants, animal-type NOS is still elusive. Recently, in pea seedlings, using the chemiluminescence assay, Corpas *et al.* [24] showed arginine-dependent NOS activity, which was constitutive, sensitive to an irreversible inhibitor of animal NOS, and dependent on the plant organ and its developmental stage.

A gene encoding NOS-like protein AtNOS1 was isolated from the *Arabidopsis* genome; it is involved in the process of growth and hormonal signaling [25]. It was also observed that AtNOS1 may function as NO source in the process of flowering control [26] and in defense response, induced by a lipopolysaccharide [27]. DNA sequencing analyses did not show affinity of AtNOS1 protein to any of animal-origin NOS isoforms. However, the most recent studies have raised critical questions regarding the nature of AtNOS1 [28, 29]. AtNOS1 (Q664P9) and the orthologous genes from rice (Q6YPG5) and maize (AY110367) have been cloned; however, after purification of recombinant protein, no NOS activity has been detected [28]. Moreover, AtNOS1 was identified as a member of GTP-binding family. On the basis of a report by Morimoto *et al.* [30], in which bacterial protein Yqett, an orthologue of AtNOS1, is defined as GTPase, it has been suggested that AtNOS1 might serve as GTPase, involved in mitochondrial ribosome biogenesis and/or processes of translation [28], and in this case, it might indirectly affect NO synthesis. Later on, it was proposed that the AtNOS1 gene be renamed as AtNOA1 – nitric oxide associated 1 [29]. Although the nature of AtNOA1 remains elusive and controversial [28, 29], there is no doubt that the identification of AtNOA1 protein and the *Atnoa1* mutant has provided an effective way to genetically control *in vivo* NOS activity and the endogenous NO levels as the *Atnoa1* mutant has been consistently shown to have impaired *in vivo* NOS activity and reduced endogenous NO levels [25–27].

Nitric oxide can also be produced by other enzymes, apart from NOS, such as NR. NO generation via NR was demonstrated *in vitro* [31] and *in vivo* [32]. NR synthesized this molecule from NO_2^-, by the participation of NAD(P)H [33]. Transformation of NO_2^- to NO occurs most probably on a molybdenum cofactor. This synthesis strictly depended on nitrite and nitrate content in the tissue [32, 34, 35]. At a high *in vitro* nitrite concentration (e.g., 100 μM), NO synthesis constituted approximately 1% of the total NR reduction activity, whereas *in vivo* NO generation was estimated at 0.01–0.1% of the NR activity [32]. NO immediately reacts with $O_2^{\bullet-}$, forming peroxynitrite that contributes to a decrease in assayed NO concentration. Taking into consideration NO loss by the value of NO reaction with $O_2^{\bullet-}$, it was shown that the production of this signaling molecule in leaves of vetch, Chinese rose, and *Arabidopsis thaliana* is almost 20 times higher than that assayed previously [36]. NO production, depending on NR activity, was also recorded in many other plant species, for example, in cucumber [37], sunflower, spinach, maize [32], *Arabidopsis* [38], wheat, orchid, aloe [39], and tobacco [40, 41] as well as in *Chlamydomonas reinhardtii* [42].

Xanthine oxidoreductase (XOR) is another Moco-containing enzyme that has been found to produce NO in plants as well as in animals. Xanthine oxidoreductase occurs in two interconvertible forms: the superoxide producing xanthine oxidase and xanthine dehydrogenase [43]. XOR has been found in pea leaf peroxisomes where the preponderant form of the enzyme is xanthine oxidase and only 30% is present in the form of xanthine dehydrogenase [44, 45]. XOR can produce the free radicals $O_2^{\bullet-}$ and $NO^{\bullet}$ during its catalytic reaction, depending on whether the oxygen tensions are high and low, respectively [46–48]. This property of producing $O_2^{\bullet-}$ and $NO^{\bullet}$ radicals confers a key role on XOR as a source of signal molecule in plant cells [49].

Another enzyme that can generate NO from nitrite is a plasma membrane-bound enzyme of tobacco roots (Ni-NOR) [50]. This enzyme has a higher molecular weight than nitrate reductase, but has to be characterized. Other good candidates for enzymatic generation of NO include horseradish peroxides [51], cytochrome P450 [52], catalase, and hemoglobin [53]. The production of NO and citrulline by horseradish peroxidase from *N*-hydroxy-arginine (NOHA) and H_2O_2 was reported a decade ago [52]. More recently, horseradish peroxidase was also demonstrated to generate NO from hydroxyurea and H_2O_2 [51]. This source of NO should be carefully considered taking into account that peroxidases are widespread enzymes, involved in important physiological processes of plant cells [54].

Heme proteins that have been proposed as good candidates for the enzymatic generation of NO are cytochromes P450. These proteins are present in plants as well as in animal systems and have been shown to catalyze the oxidation of NOHA by NADPH and O_2 with the generation of NO [53, 55]. Hemoglobin and catalase were also reported to produce NO and other nitrogen oxides by catalyzing the oxidation of NOHA by cumyl hydroperoxide [52].

In plants, nitric oxide can also be generated by nonenzymatic mechanisms. Nitrification/denitrification cycle provides NO as a by-product of N_2O oxidation into the atmosphere [21]. It is known that the nonenzymatic reduction of nitrite can lead to the formation of NO, and this reaction is favored at acidic pH when nitrite can dismutate the NO and nitrate [9]. Nitrite can also be chemically reduced by ascorbic acid at pH 3–6 to yield NO and dehydroascorbic acid [56]. This reaction could occur under microlocalized pH conditions in the chloroplast and apoplastic space where ascorbic acid is known to be present [57]. In barley aleurone layer cells, NO can also be synthesized by the reduction of nitrite by ascorbate at acidic pH [58]. Another nonenzymatic mechanism proposed for NO formation is the light-mediated reduction of NO_2 by carotenoids [59].

1.4 Physiological Role of Nitric Oxide

Nitric oxide has emerged as an important signaling molecule associated with many biochemical and physiological processes in plants [60–62]. NO was classified as a phytohormone that might function as a gaseous endogenous plant growth regulator [63] as well as a nontraditional plant growth regulator [16]. It has the capability to regulate diverse physiological processes in a concentration-dependent manner [64, 65], such as root organogenesis, hypocotyl growth, defense responses, stomatal movement, apoptosis, hypersensitive responses, growth and development, and phytoalaxin production [10, 19, 60, 66–72], under different environmental conditions. Therefore, in recent years, the role of NO in regulating various physiological and biochemical activities in plants has become an important area of research. In this section, we only discuss the role of NO in different processes of plants under normal conditions (unstressed plants) because the role of NO in plants under different abiotic and biotic stresses is discussed in other chapters.

1.4.1 Effect of Nitric Oxide on Seed Dormancy

Dormancy prevents seed germination under conditions that would otherwise allow germination. Many endogenous compounds reduce/break seed dormancy; among them are nitrogen-containing compounds that include nitrate, nitrite, hydroxylamine, azide, and sodium nitroprusside (SNP). The ability of SNP to reduce seed dormancy in lettuce [72], *Arabidopsis* [73–75], and barley [74] led to the conclusion that NO played some role in seed germination. Moreover, the stimulatory effect of NO on seed germination has also been reported in other crops.

NO stimulated seed germination in *Paulonia tomentosa* [76] under normal conditions as well as in *Suaeda salsa* under NaCl stress [77]. SNP (up to 0.8 mM) application promoted seed germination in lupin that was more pronounced after 18 and 24 h and ceased after 48 h [78]. Similarly in canola, NO stimulated seed germination in a dose-dependent manner: lower concentrations of SNP (0.05–0.5 mM) enhanced seed germination up to 18 h, whereas high concentrations (1 and 2 mM) inhibited the process [79]. Furthermore, exogenous application of nitric oxide also promoted seed germination in maize [80].

1.4.2
Effect of Nitric Oxide on Growth

In rapidly growing pea seedlings, treatment of NO showed a dual behavior: lower concentrations (micromolar) increased the rate of leaf expansion, but no beneficial effect was noticed [81] at higher concentrations. Similarly, high concentrations of NO (40–80 ppm) inhibited the growth of tomato whereas low concentrations (0–20 ppm) stimulated the growth of tomato, lettuce [82], and pea seedling [81]. NO also activated the growth of root segments of maize comparable to that by indole acetic acid [65]. Although SNP (0.1 mM) inhibited growth of hypocotyls in potato, lettuce, and *Arabidopsis* [72], it induced root development in cucumber [60]. Exogenous application of NO inhibited the elongation of mesocotyl in maize seedling [83]. Contrary to this, an increase in the leaf biomass of maize seedlings was observed by the endogenously produced and exogenously applied NO [84]. The effect of NO on plant growth was found to be concentration dependent [64, 65]. Treating maize seedlings with lower concentration of SNP promoted root growth whereas higher concentration was inhibitory. Seedlings of canola, raised from the seeds treated with lower concentration of SNP, had more root length and dry mass whereas higher concentration reduced the values of these parameters [79]. A similar dual behavior of NO donor SNP was also noted in wheat [85].

1.4.3
Effect of Nitric Oxide on Senescence

Senescence is a process characterized by water loss and desiccation of plant tissues. Some studies suggest that NO has antisenescence properties. Exogenous application of NO in pea leaves under senescence promoting conditions decreased ethylene level because of the inhibition of ethylene biosynthesis [13, 63, 81]. In *Arabidopsis*, however, the level of ethylene increased significantly after being exposed to NO gas [71]. It was also observed that NO emission decreased as ethylene production increased from anthesis to senescence [78]. NO donors exert a protective effect against abscisic acid (ABA)-induced senescence of rice leaves by diminishing ABA-dependent effects such as leaf senescence, enhanced H_2O_2 and malondialdehyde

(MDA) content, reduction in GSH, ascorbic acid level, and antioxidant enzyme activity [86]. The protective effect was reversed by NO scavenger (PTIO) suggesting that the observed phenomenon may be attributed to NO.

1.4.4 Effect of Nitric Oxide on Nitrate Reductase Activity

Nitrate reductase activity is one of the NO sources in plant roots. Exogenous application of SNP (100 μM) significantly enhanced the activity of nitrate reductase in leaves of maize plants [80]; however, in the roots of pea and wheat, SNP did not influence NR activity [87].

1.4.5 Effect of Nitric Oxide on Respiration

NO affects the mitochondrial functionality in plant cells and reduces total cell respiration due to its inhibitory effect on the cytochrome functioning. In carrot cell suspension, NO reduced total respiration by 50%, and this effect was accompanied by a significant increase in cell death. Similarly, in soybean cotyledon mitochondria, the oxygen uptake was inhibited after NO treatment, but it was restored upon NO depletion [88]. It was concluded that alternative oxidase (AOX) may play a role in NO tolerance in higher plants. Nitric oxide can also modulate other mitochondrial enzymes, tobacco aconitase, which is a constituent of Krebs cycle. Its inactivation by NO decreases the cellular energy metabolism that may result in reduced electron flow through the mitochondrial respiratory electron transport chain and a subsequent decrease in the generation of reactive oxygen species (ROS), the natural by-product of respiration [89].

1.4.6 Effect of Nitric Oxide on Stomatal Movement

NO has also been reported to play a role in stomatal movement being, together with H_2O_2, an indispensable component of ABA-induced stomatal closure [22, 38, 90, 91]. The exogenous application of NO to both monocotyledonous and dicotyledonous epidermal strips induced stomatal closure through a Ca^{2+}-dependent process [92]. In *Pisum sativum* and *Vicia faba* plants, abscisic acid increased the endogenous production of NO that was suggested to be the reason for ABA-induced stomatal closure [93]. There are also some convergent evidences that support the involvement of nitrate reductase through the production of NO in guard cells [90] leading to their closure [90, 93].

1.4.7 Effect of Nitric Oxide on Chlorophyll Content

NO donors (SNP) have been found to enhance chlorophyll content in potato, lettuce, and *Arabidopsis* [72]. The role of NO in preserving and increasing chlorophyll content in pea and potato [94] was also proved. The protective effect of NO on the chlorophyll retention may reflect NO effects on iron availability. A strong evidence supporting the role of NO in iron nutrition of plants was presented by Graziano *et al.* [95] as iron-deficient growth conditions normally result in chlorosis. NO treatment increased the chlorophyll content in maize leaves up to the control level [95].

1.4.8
Effect of Nitric Oxide on Photosynthesis

Photosynthesis is one of the most important physiological processes. The whole metabolism of plants directly or indirectly depends on this process; any change in photosynthetic rate will automatically affect the rest of the processes in plant. However, the role of NO in photosynthesis is poorly understood, which is well indicated by the modest number of *in vivo* and *in vitro* studies in this area with mixed results [96, 97]. Nitric oxide and its donors such as sodium nitroprusside, *S*-nitroso-*N*-acetylpenicillamine (SNAP), and *S*-nitrosoglutathione (GSNO) are recognized to differentially regulate the photosynthetic rate. NO gas decreases net photosynthetic rate in *Avena sativa* and *Medicago sativa* leaves [98]. NO donor SNP has been found to decrease the level of enzymes that regulate photosynthesis in wheat [99] and in *Phaseolus vulgaris* [100].

Nitric oxide is able to influence the photosynthetic electron transport chain directly. PS II is an important site for NO action [101]; within PS II complex, important binding sites of NO are the nonheme iron between Q_A and Q_B binding sites [102], Y_D, Tyr residue of D2 protein [103], and manganese (Mn) cluster of water-oxidizing complex [104].

NO donor SNAP does not modify the maximal quantum efficiency (F_v/F_m) but inhibits the linear electron transport rate and light-induced pH formation (ΔpH) across thylakoid membrane, and decreased the rate of ATP synthesis [96]. Another NO donor, sodium nitroprusside, reduces quantum efficiency (F_v/F_m) in the intact potato leaves but causes no difference in ΔpH-dependent nonphotochemical quenching (NPQ) [97]. A moderate decrease in F_v/F_m was also observed by SNP treatment in pea leaves [105]. Moreover, NO donor has also been found to slow down the electron transfer between the primary and the secondary quinone electron acceptor *in vivo*, in a concentration-dependent manner [101, 105, 106].

S-Nitrosoglutathione, another NO donor, caused a significant decrease in F_v/F_m value in intact pea leaves and decreased steady-state qP, which indicated that NO increased the proportion of closed PS II reaction center, besides reducing steady-state transient NPQ [101] that resembles reaction center NPG, described by Finazzi *et al.* [107] in *Hordeum vulgare*. Wodala *et al.* [101] suggested different chemical properties of NO donors and different experimental conditions as the reasons to account for the above conflicting results.

1.4.9
Effect of Nitric Oxide on Antioxidant System

It is now a common belief that NO acts as a second messenger in plants. One of the most intriguing issues in NO biology is its dual function as a potent oxidant and effective antioxidant [108]. This dual role of NO might depend on its concentration as well as on the status of the environment. Oxidative stress is the common result of the action of many environmental factors, manifesting itself in a cell by an increased level of reactive oxygen species [109]. The cytoprotective role of NO is mainly based on its ability to maintain the cellular redox homeostasis and to regulate the level and toxicity of ROS.

The ability of NO to exert a protective function against oxidative stress is caused by the factors such as

(a) reaction with lipid radicals, which stops the propagation of lipid oxidation;
(b) scavenging the superoxide anion and formation of peroxynitrite ($ONOO^-$) that is toxic for plants but can be neutralized by ascorbate and glutathione;
(c) activation of antioxidant enzymes (SOD, CAT, and POX).

One of the fastest reactions of NO within a biological system is its combination with superoxide anion ($O_2^{\bullet -}$) that leads to the formation of strong oxidant peroxynitrite ($ONOO^-$) [10, 19] that is one of the major toxic reactive nitrogen species [20] that exerts deleterious effects on DNA, lipids, and proteins [20, 35, 110].

The effect of NO on peroxidase is still scarce and somewhat controvertible; the lower concentration of NO donor SNP increases peroxidase activity in *Brassica* whereas higher concentration proved inhibitory [79]. Similarly, ascorbate peroxidase activity was inhibited by higher SNP concentration in tobacco and canola [111]. Moreover, higher concentration of SNP inhibited coniferyl alcohol peroxidase activity in *Zinnia elegna* [112].

Treatment of wheat plant with lower concentration of SNP decreased H_2O_2 content, but antioxidant activity was enhanced [85]. Moreover, NO can react with lipid alcoxy ($LO^{\bullet}$) and peroxyl ($LOO^{\bullet}$) radicals to stop the propagation of radical-mediated lipid oxidation in a direct fashion [85, 113]. NO decreased TBARS content in wheat seedlings [85].

1.4.10
Effect of Nitric Oxide on Programmed Cell Death

There are contradictory reports concerning NO and programmed cell death (PCD). The elevated levels of NO were sufficient to induce cell death in *Arabidopsis* cell suspension, independent of reactive oxygen species [111]. An increase in either NO or ROS individually did not induce cell death, whereas simultaneous increase in NO and ROS activated the process of cell death with typical cytological and biochemical features of PCD [114]. Moreover, the interaction between NO and ROS in PCD induction was also investigated in soybean cell suspension [115], and the researchers concluded that NO by itself does not induce PCD, but the key factor determining it is the NO:superoxide ratio [115]. Contrary to that in *Taxus brevifolia* and *Kalanchoe diagremontiana*, SNP caused nitric oxide burst, which led to a significant increase in nuclear DNA fragmentation and cell death [116]. On the other hand, it was suggested that NO donors delay PCD in barley aleurone layers treated with GA, but do not inhibit metabolism in general or the GA-induced synthesis and secretion of α-amylase. α-Amylase secretion is stimulated slightly by NO donor. The effect of NO donors is specific for NO because they can be blocked completely by the NO scavenger 2-(4-carboxyphenyl)-4,4,5,5-tetramethylimidazoline-1-oxyl-3-oxide. Thus, NO may be an endogenous modulator of PCD in barley aleurone cell.

1.5 Nitric Oxide and Cross Talk with Classical Plant Hormones

In this section, we are going to discuss the cross talk between NO and recognized hormones that act simultaneously in different physiological and biochemical processes in plants.

1.5.1 Auxins and Nitric Oxide

NO induced the elongation of maize root segments in a dose-dependent manner [65]. It has, therefore, been proposed that the auxin indole acetic acid (IAA) and NO might share some common steps in the signal transduction pathway because both elicit the same responses in plants. The dependence of auxin on NO in the induction of adventitious root development was recently demonstrated in cucumber explants [60]. Moreover, explants from wood species were also responsive to NO treatment to induce adventitious root formation [117].

In cucumber explants, IAA treatment induces a transient increase in the level of endogenous NO in the basal region of the hypocotyl, where the new meristem develops [60]. This localized NO bulk might stimulate the GC-catalyzed synthesis of cGMP [118]. The GC inhibitor reduced adventitious root formation in both IAA- and NO-treated cucumber explants. This effect was, however, reversed when permeable cGMP analogue was added together with GC inhibitor and NO or IAA [60].

Earlier in tobacco, activation of defense genes by NO was also induced by cGMP [68]. These genes may act via cADPR that, in turn, regulates Ca^{2+} level in plants [119]. Variations in [Ca^{2+}] might play a role in the signal transduction pathway leading to the activation of the process of mitotic differentiation to initiate rooting.

Nitric oxide can also act via a cGMP-independent pathway, activating phosphatases and protein kinases including MAPKs. Interestingly, a rapid and transient increase in MAPK activity in response to low level of auxins was reported in *Arabidopsis* seedling roots [120].

IAA-induced endogenous NO bulk in roots can result in a bifurcated signal transduction pathway in which NO mediates a cGMP-dependent or -independent increase in cytosolic Ca^{2+}, which in turn triggers changes in plant gene expression leading to the auxin response.

1.5.2
Abscisic Acid and Nitric Oxide

ABA regulates various vital processes in plants where stomatal movement is one of them. In the guard cell, ABA induces the depolarization of the plasma membrane potential that leads to the generation of a driving force for K^+ efflux, inactivates K^+ influx through inward-rectifying K^+ (K^+_{in}) channels, and activates a current through outward-rectifying (K^+_{out}) channels. These changes together with both slow and fast activating anion channels facilitate the net loss of salt from the cell [121]. Both cytosolic free Ca^{2+} concentration (Ca^{2+}_{cyt}) and cytosolic pH have been reported to participate as second messengers of this response [122]. ABA induces guard cell $[Ca^{2+}]_{cyt}$ elevation either by influx from extracellular space or by release from internal source [123], which leads to the loss of guard cell turgor, favoring stomatal closer.

On the other hand, an exogenous application of NO to both monocotyledonous and dicotyledonous epidermis strips was sufficient to induce stomatal closure through a Ca^{2+}-dependent process [92]. Moreover, it was also reported that in *P. sativum* and *V. faba*, ABA induces an increase in endogenous NO level. This bulk of ABA-induced NO production was reported to be sufficient and necessary for ABA induction of stomatal closure [90, 124]. The participation of NO as a signal molecule in guard cell movement is a very recent topic, and much work is still to be done in this field.

1.5.3
Cytokinins, Gibberellins, and Nitric Oxide

Cytokinins (CKs) stimulate photomorphogenic responses, mainly those related to the de-etiolation and pigment synthesis [125]. In exogenous application, CKs have been reported to inhibit hypocotyl elongation in seedlings grown in the dark [126]. Similarly, NO reduces hypocotyl elongation in *Arabidopsis* and lettuce seedlings grown in the dark [72]. Moreover, it had also been reported that the CK treatment to cotyledons and leaves grown under dark condition could not cause etiolation to revert completely. However, CK had an ability to abolish the lag phase because in chlorophyll production during subsequent illumination it abolishes the lag phase [127]. On the other hand, NO has also been reported to slightly increase the chlorophyll level in wheat seedlings grown in the dark [72]. Thus, the effect of NO is similar to that of CKs.

As mentioned earlier, CKs regulate the synthesis of some pigments such as anthocyanins and betacyanins. NO plays the same role as CK action on betacyanin accumulation. Moreover, NOS inhibitor and an NO scavenger blocked the action of CKs on betacyanin accumulation [128], suggesting that NO somehow helps CK to promote that response or NO is necessary to accomplish CK function. The first evidence suggesting a direct relationship between CKs and NO production was that the exogenous application of CKs to *Arabidopsis*, parsley, or tobacco cell cultures leads to a rapid stimulation of NO release [129].

Some seeds require light for their germination under certain conditions. In such cases, GA has been found to act as an active form of phytochrome [130] to induce germination. However, CK alone is generally ineffective in breaking dormancy, but when it acts synergistically with light or GAs, it allows germination [131]. The germination of lettuce seed (cv. Grand Rapid) is also a phytochrome-dependent process, and it was observed that NO donors are able to stimulate germination in the dark, similar to GA, or under a few minute pulse of white light [72]. However, seeds were also able to germinate in light in the presence of NO scavenger, suggesting that light and NO can stimulate germination in different ways [72]. Moreover, it is yet to be determined whether GA and NO promote germination through the same or different pathways.

1.5.4 Ethylene and Nitric Oxide

Ethylene plays an active role in many plant responses [132]. It was suggested that NO and ethylene caused an antagonistic effect during maturation and senescence of the plant [13]. It was demonstrated that endogenous NO and ethylene content maintain an inverse correlation during the ripening of strawberries and avocados [133] while unripe, green fruits contain high NO and low ethylene concentrations; the maturation process is accompanied by a marked decrease in NO concomitant with an increase in ethylene [133].

References

1 Schmidt, H.H.H.W. and Walter, U. (1994) *Cell*, **78**, 919–925.

2 Koshland, D.E., Jr. (1992) *Science*, **258**, 1861.

3 Klepper, L.A. (1975) *WSSA Abstr.*, **184**, 70.

4 Klepper, L.A. (1978) *Plant Physiol.*, **61** (Suppl.), 65.

5 Klepper, L.A. (1979) *Atmos. Environ.*, **13**, 537–542.

6 Klepper, L.A. (1990) *Plant Physiol.*, **93**, 26–32.

7 Klepper, L.A. (1991) *Pest Biochem. Physiol.*, **39**, 43–48.

8 Klepper, L.A. (1987) *Plant Physiol.*, **85**, 96–99.

9 Stohr, C. and Ullrich, W.R. (2002) *J. Exp. Bot.*, **53**, 2293–2303.

10 Neill, S.J., Desikan, R., and Hancock, J.T. (2003) *New Phytol.*, **159**, 11–35.

11 Graziano, M. and Lamattina, L. (2005) *Trends Plant Sci.*, **10**, 4–8.

12 Davies, P.J. (1995) The plant hormone concept: concentration, sensitivity and transport, in *Plant Hormones: Physiology, Biochemistry and Molecular Biology* (ed. P.J. Davies), Kluwer Academic Press, Dordrecht, The Netherlands, pp. 13–18.

13 Leshem, Y.Y., Wills, R.B.H., and Ku, V.V.V. (1998) *Plant Physiol. Biochem.*, **36**, 825–833.

14 Caro, A. and Puntarulo, S. (1999) *Free Radic. Res.*, **31**, 205–212.

15 Trewaras, A.J. and Malho, R. (1997) *Plant Cell*, **9**, 1181–1195.

16 Beligni, M.V. and Lamattina, L. (2001) *Trends Plant Sci.*, **6**, 508–509.

17 Leshem, Y.Y. (1996) *Plant Growth Regul.*, **18**, 155–169.

18 Arasimowicz, M. and Wieczorek, J.F. (2007) *Plant Sci.*, **172**, 876–887.

19 Wendehenne, D., Pugin, A., Klessig, D.F., and Durner, J. (2001) *Trends Plants Sci.*, **6**, 177–183.

20 Stamler, J.S., Singel, D.J., and Loscalzo, J. (1992) *Science*, **258**, 1898–1902.

21 Wojtaszek, P. (2000) *Phytochemistry*, **54**, 1–4.

22 Garcia-Mata, C. and Lamattina, L. (2003) *Trends Plant Sci.*, **8**, 20–26.

23 Stamler, J.S. (1994) *Cell*, **78**, 931–936.

24 Corpas, F.J., Barroso, J.B., Carreras, A., Valderrama, R., Palma, J.M., Leon, A.M., Sandalio, L.M., and del Rio, L.A. (2006) *Planta*, **224**, 246–254.

25 Guo, F., Okamoto, M., and Crawford, N.M. (2003) *Science*, **302**, 100–103.

26 He, Y., Tang, R.H., Hao, Y., Stevens, R., Cook, C.W., Ahn, S.M., Jing, L., Yang, Z., Chen, L., Guo, F., Fiovani, F., Jackson, R.B., Crawford, N.M., and Pei, Z.M. (2004) *Science*, **305**, 1968–1971.

27 Zeidler, D., Zahringer, U., Gerber, I., Dubery, I., Hartung, T., Bors, W., Hutzler, P., and Durner, J. (2004) *Proc. Natl. Acad. Sci. USA*, **101**, 15811–15816.

28 Zemojtel, T., Froblich, A., Palmicri, M.C., Kolanczyk, M., Mikula, I., and Wyrwicz, L.S. (2006) *Trends Plant Sci.*, **11**, 524–525.

29 Crawford, N.M., Galli, M., Tischner, R., Heimer, Y.M., Okamobo, M., and Mack, A. (2006) *Trends Plant Sci.*, **11**, 526–527.

30 Morimoto, T., Loh, P.C., Hirai, T., Asiai, K., Kobayashi, K., Moriya, S., and Ogasawara, N. (2002) *Microbiology*, **148**, 3539–3552.

31 Yamasaki, H. and Sakihama, Y. (2000) *FEBS Lett.*, **468**, 89–92.

32 Rockel, P., Strube, F., Rockel, A., Wildt, J., and Kaiser, W.M. (2002) *J. Exp. Bot.*, **53**, 1–8.

33 Kaiser, W.M., Weiner, H., Kandlbinder, A., Tsai, C.B., Rockel, P., Sonoda, M., and Planchet, E. (2002) *J. Exp. Bot.*, **53**, 875–882.

34 Kaiser, W.M. and Huber, S.C. (2001) *J. Exp. Bot.*, **52**, 1981–1989.

35 Yamasaki, H., Sakihama, Y., and Takahashi, S. (1999) *Trends Plant Sci.*, **4**, 128–129.

36 Vanin, A., Svistunenko, D.A., Mikoyan, V.D., Serezhenkov, V.A., Fryer, M.J., Baker, N.R., and Cooper, C.E. (2004) *J. Biol. Chem.*, **279**, 24100–24107.

37 De la Haba, P., Aguera, E., Benitez, L., and Maldonado, J.M. (2001) *Plant Sci.*, **16**, 231–237.

38 Desikan, R., Griffiths, R., Hancock, J., and Neill, S. (2002) *Proc. Natl. Acad. Sci. USA*, **99**, 16314–16318.

39 Xu, Y.C. and Zhao, B.L. (2003) *Plant Physiol. Biochem.*, **41**, 833–838.

40 Planchet, F., Kapuganti, J.G., Sonoda, M., and Kaiser, W.M. (2005) *Plant J.*, **41**, 732–743.

41 Planchet, E., Sonoda, M., Zeier, J., and Kaiser, W.M. (2006) *Plant Cell Environ.*, **29**, 59–69.

42 Sakihama, Y., Nakamura, S., and Yamasaki, H. (2002) *Plant Cell Physiol.*, **43**, 290–297.

43 Palma, J.M., Sandalio, L.M., Corpas, F.J., Romero-Puertas, M.C., McCarthy, I., and del Rio, L.A. (2002) *Plant Physiol. Biochem.*, **40**, 521–530.

44 Corpas, F.J., de la Colina, C., Sanchez-Rosevo, F., and del Rio, L.A. (1997) *J. Plant Physiol.*, **151**, 246–250.

45 Sandalio, L.M. and del Rio, L.A. (1988) *Plant Physiol.*, **88**, 1215–1218.

46 Millar, T.M., Stevens, C.R., Benjamin, N., Eisenthal, R., Harrison, R., and Blacke, D.R. (1998) *FEBS Lett.*, **427**, 225–228.

47 Godber, B.L.J., Doel, J.J., Sapkota, G.P., Blake, D.R., Stevens, C.R., Eisenthal, R., and Harrison, R. (2000) *J. Biol. Chem.*, **275**, 7757–7763.

48 Harrison, R. (2002) *Free Radic. Biol. Med.*, **33**, 774–797.

49 Corpas, F.J., Barroso, J.B., and del Rio, L.A. (2001) *Trends Plant Sci.*, **6**, 145–150.

50 Stohr, C., Strule, F., Marx, G., Ullrich, W.R., and Rockel, P. (2001) *Planta*, **212**, 835–841.

51 Huang, X., Von Rad, U., and Durner, J. (2002) *Planta*, **215**, 914–923.

52 Boucher, J.L., Genet, A., Vadon, S., Delaforge, M., Henry, Y., and Mansuy, D. (1992) *Biochem. Biophys. Res. Commun.*, **187**, 880–886.

53 Boucher, J.L., Genet, A., Vadon, S., Delaforge, M., Henry, Y., and Mansuy, D. (1992) *Biochem. Biophys. Res. Commun.*, **184**, 1158–1164.

54 Veitch, N.C. (2004) *Phytochemistry*, **65**, 249–259.

55 Mansuy, D. and Boucher, J.L. (2002) *Drug Metab. Rev.*, **34**, 593–606.

56 Henry, Y.A., Ducastel, B., and Guissani, A. (1997) *Nitric Oxide Research from Chemistry to Biology* (eds Y.A. Henry, A. Guissani, and B. Ducastel), Landes Company, pp. 15–46.

57 Horemans, N., Foyer, C.H., and Asard, H. (2000) *Trends Plant Sci.*, **5**, 263–267.

58 Beligni, M.V., Fath, A., Bethke, P.C., Lamattina, L., and Jones, R.L. (2002) *Plant Physiol.*, **129**, 1649–1650.

59 Cooney, R.V., Harwood, P.T., Custer, L., and Franke, A.A. (1994) *Environ. Health Perspect.*, **102**, 460–462.

60 Pagnussat, G., Simontachi, M., Puntarulo, S., and Lamattina, L. (2002) *Plant Physiol.*, **129**, 954–956.

61 Lamattina, L., Garcia-Mata, C., Graziano, M., and Pagnussat, G. (2003) *Annu. Rev. Plant Biol.*, **54**, 109–136.

62 Stohr, C. and Stremlau, S. (2006) *J. Exp. Bot.*, **57**, 463–470.

63 Leshem, Y.Y. (2000) Nitric oxide in plants, in *Occurrence, Function and Use*, Kluwer Academic Publishers, Dordrecht, The Netherlands.

64 Anderson, L. and Mansfield, T.A. (1979) *Environ. Pollut.*, **20**, 113–121.

65 Gouvea, C.M.C.P., Souza, J.F., Magalhaes, C.A.N., and Martins, I.J. (1997) *Plant Growth Regul.*, **21**, 183–187.

66 Noritake, T., Kawakita, K., and Doke, N. (1996) *Plant Cell Physiol.*, **37**, 113–116.

67 Delledonne, M., Xia, Y., Dixon, R.A., and Lamb, C. (1998) *Nature*, **394**, 585–588.

68 Durner, J., Wendehenne, D., and Klessig, F. (1998) *Proc. Natl. Acad. Sci. USA*, **9**, 10328–10333.

69 Kim, O.K., Murakami, A., Nakamura, Y., and Ohigashi, H. (1998) *Cancer Lett.*, **125**, 199–207.

70 Durner, J. and Klessig, D.F. (1999) *Curr. Opin. Plant Biol.*, **2**, 369–374.

71 Magalhaes, J.R., Monte, D.C., and Durzan, D. (2000) *Physiol. Mol. Biol. Plants*, **6**, 117–127.

72 Beligni, M.V. and Lamattina, L. (2000) *Planta*, **210**, 215–221.

73 Batak, L., Devic, M., Giba, Z., Grubisic, A., Poff, K.L., and Konjevic, R. (2002) *Seed Sci. Res.*, **12**, 253–257.

74 Bethke, P.C., Gubler, F., Jacobsen, J.V., and Jones, R.L. (2004) *Planta*, **219**, 847–855.

75 Bethke, P.C., Libourel, I.G.L., and Jones, L.R. (2006) *J. Exp. Bot.*, **57**, 517–526.

76 Giba, Z., Grubuisic, D., Todorovic, S., Saje, L., Stojakovic, D., and Konjevic, R. (1998) *Plant Growth Regul.*, **26**, 175–181.

77 Li, W.-Q., Liu, X.J., Khan, M.A., and Yamaguchi, S. (2005) *J. Plant Res.*, **118**, 207–214.

78 Kopyra, M. and Gwozdz, E.A. (2003) *Plant Physiol. Biochem.*, **41**, 1011–1017.

79 Zanardo, D.I.L., Zanardo, F.M.L., Ferrarese, M.D.L.L., Magalhaes, J.R., and Filho, O.F. (2005) *Physiol. Mol. Biol. Plants*, **11**, 81–86.

80 Zhang-Shao, Y., Ren-Xiao, L., and Cheng-Shun, C. (2004) *Plant Physiol. Commun.*, **40**, 309–310.

81 Leshem, Y.Y. and Haramaty, E. (1996) *J. Plant Physiol.*, **148**, 258–263.

82 Hofton, C.A., Besford, P.T., and Wellburn, A.R. (1996) *New Phytol.*, **133**, 495–501.

83 Zhang, M., An, L., Feng, H., Chen, T., Chen, K., Liu, Y., Tang, H., Chang, J., and Wang, X. (2003) *Photochem. Photobiol.*, **77**, 219–225.

84 An, L., Liu, Y., Zhang, M., Chen, T., and Wang, X. (2005) *J. Plant Physiol.*, **162**, 317–326.

85 Tian, X. and Lei, Y. (2006) *Biol. Plant.*, **50**, 775–778.

86 Hung, K.T. and Kao, C.H. (2003) *J. Plant Physiol.*, **160**, 871–879.

87 Kolbert, Z., Bartha, B., and Erdei, L. (2005) *Acta Biol. Szeg.*, **49**, 13–16.

88 Millar, A.H. and Day, D.A. (1996) *FEBS Lett.*, **398**, 155–158.
89 Navarre, D., Wendenhenne, D., Durner, J., Noad, R., and Klessing, D.F. (2000) *Plant Physiol.*, **122**, 573–582.
90 Garcia-Mata, C. and Lamattina, L. (2002) *Plant Physiol.*, **128**, 790–792.
91 Desikan, R., Cheung, M.K., Bright, J., Henson, D., Hancock, J.T., and Neill, S.J. (2004) *J. Exp. Bot.*, **55**, 205–212.
92 Garcia-Mata, C. and Lamattina, L. (2001) *Plant Physiol.*, **126**, 1196–1204.
93 Neill, S.J., Desikan, R., Clarke, A., and Hancock, J.T. (2002) *Plant Physiol.*, **128**, 13–16.
94 Leshem, Y.Y., Haramaty, E., Iiuz, D., Malik, D., Sofer, Y., Roitman, L., and Leshem, Y. (1997) *Plant Physiol. Biochem.*, **35**, 573–579.
95 Graziano, M., Beligni, M.V., and Lamattina, L. (2002) *Plant Physiol.*, **130**, 1852–1859.
96 Takahashi, S. and Yamasaki, H. (2002) *FEBS Lett.*, **512**, 145–148.
97 Yang, J.D., Zhao, H.L., Zhang, T.H., and Yun, J.F. (2004) *Acta Bot. Sin.*, **46**, 1009–1014.
98 Hill, A.C. and Bennett, J.H. (1970) *Atoms Environ.*, **4**, 341–348.
99 Tu, J., Shen, W.B., and Xu, L.L. (2003) *Acta Bot. Sin.*, **45**, 1055–1062.
100 Lum, H.K., Lee, C.H., Butt, Y.K.C., and Lo, S.C.L. (2005) *Nitric Oxide*, **12**, 220–230.
101 Wodala, B., Deak, Z., Vass, I., Erdei, L., Altorjay, I., and Horvath, F. (2008) *Plant Physiol.*, **146**, 1920–1927.
102 Diner, B.A. and Petrouleas, V. (1990) *Biochem. Biophys. Acta*, **1015**, 141–149.
103 Sanakis, Y., Goussias, C., Mason, R.P., and Petrouleas, V. (1997) *Biochemistry*, **36**, 1411–1417.
104 Schansker, G., Goussias, C., Petrouleas, V., and Rutherford, A.W. (2002) *Biochemistry*, **41**, 3057–3064.
105 Wodala, B. (2006) *Acta Physiol. Szeg.*, **50**, 185–188.
106 Petrouleas, V. and Diner, B.A. (1990) *Biochem. Biophys. Acta*, **1015**, 131–140.
107 Finazzi, G., Johnson, G.N., Dall'osto, L., Joliot, P., Wollman, F.A., and Bassi, R. (2004) *Proc. Natl. Acad. Sci. USA*, **101**, 12375–12380.
108 Beligni, M.V. and Lamattina, L. (1999) *Trends Plant Sci.*, **4**, 299–300.
109 Mittler, R. (2002) *Trends Plant Sci.*, **7**, 405–410.
110 Pryor, W.A. and Squadrito, G.L. (1995) *Am. J. Physiol.*, **268**, L699–L700.
111 Clark, D., Durner, J., Navarre, D.A., and Klessig, D.F. (2000) *Mol. Plant Microb. Interact.*, **13**, 1380–1384.
112 Ferrer, M.A. and Ros-Barcelo, A. (1999) *Plant Cell Environ.*, **22**, 891–897.
113 Lamotte, O., Gould, K., Lecourieux, D., Sequeira-Legrand, A., Lebun-Garcia, A., Durner, J., Pugin, A., and Wendehenne, D. (2004) *Plant Physiol.*, **135**, 516–529.
114 de-Pinto, M.C., Tomassi, F., and de Gara, L. (2002) *Plant Physiol.*, **130**, 689–708.
115 Delledonne, M., Zeier, J., Marocco, A., and Lamb, C. (2001) *Proc. Natl. Acad. Sci. USA*, **98**, 13454–13459.
116 Pedroso, M.C., Magalhaes, J.R., and Durzan, D. (2000) *J. Exp. Bot.*, **51**, 1027–1103.
117 Lamattina, L., Beligni, M.V., Gracia-Mata, C., and Laxalt, A.M. (2001) US Patent 6,242,384.
118 McDonald, L.J. and Murad, F. (1995) *Adv. Pharmacol.*, **34**, 263–276.
119 Leckie, C.P., McAisnish, M.R., and Hetherington, A.M. (1998) *Plant Physiol.*, **114**, 270–272.
120 Pfeiffer, S., Janistyn, B., Jessner, G., Pichorner, H., and Ebermann, R.E. (1994) *Phytochemistry*, **36**, 259–262.
121 Blatt, M.R. (2000) *Annu. Rev. Cell Dev. Biol.*, **16**, 221–241.
122 Blatt, M. and Grabov, A. (1997) *Physiol. Plant.*, **100**, 481–490.
123 Schroeder, J.I., Kwak, J.M., and Allen, G.J. (2001) *Nature*, **410**, 327–330.
124 Neill, S.J., Desikan, R., Clarke, A., Hurst, R.D., and Hancock, J.T. (2002) *J. Exp. Bot.*, **53**, 1237–1247.

125 Thomas, T.H., Hare, P.D., and van Staden, J. (1997) *Plant Growth Regul.*, **23**, 105–122.

126 Chory, J., Reinecke, D., Sim, S., Washburn, T., and Brenner, M. (1994) *Plant Physiol.*, **104**, 339–347.

127 Dei, M. (1982) *Physiol. Plant.*, **64**, 153–160.

128 Scherer, G.F.E. and Holk, A. (2000) *Plant Growth Regul.*, **32**, 345–350.

129 Tun, N.N., Holk, A., and Scherer, F.E. (2001) *FEBS Lett.*, **509**, 174–176.

130 Yang, Y.Y., Nagatani, A., Zhao, Y.J., Kang, B.J., Kendrick, B.E., and Kamiya, Y. (1995) *Plant Cell Physiol.*, **36**, 1205–1211.

131 Thomas, T.H. (1984) *Plant Growth Regul.*, **11**, 239–248.

132 Abeles, F.B., Morgan, P.W., and Saltveit, M.E. (1992) *Ethylene in Plant Biology*, Academic Press, San Diego, CA.

133 Leshem, Y.Y. and Pinchasov, Y. (2002) *J. Exp. Bot.*, **51**, 1471–1473.

2
Electron Paramagnetic Resonance as a Tool to Study Nitric Oxide Generation in Plants

Susana Puntarulo, Sebastián Jasid, Alejandro D. Boveris, and Marcela Simontacchi

Summary

Plants produce and release nitric oxide (NO) during certain physiological processes and especially under stressful conditions, primarily in actively growing tissues such as embryonic axes. A distinctive EPR signal for the adduct $(MGD)_2$-Fe^{2+}-NO ($g = 2.03$ and $a_N = 12.5$ G) was detected in axes isolated from sorghum and soybean seeds. *In vivo* generation of NO in plants is achieved through different enzymatic pathways employing either nitrite or arginine as substrates. Since NO was identified as a product from nitrate reductase (NR) activity, homogenates from axes were supplemented with $(MGD)_2$-Fe^{2+} and substrates for NADH-dependent NR activity, and the rate of generation of NO was measured by EPR to assess maximal on line NO generation by enzymatic activity. The $(MGD)_2$-Fe^{2+}-NO spectrum was recorded for 10 min and the increase in peak height at 3412 G was plotted. A similar assay was developed for assessing nitric oxide synthase (NOS)-like activity in embryonic axes. Thus, EPR was successfully employed to identify NO presence upon germination and to assess NR and NOS-like activity by direct identification of the generated product, NO. The main advantages of EPR methods, such as good potential specificity and reasonable time resolution, overcome the poor sensitivity that could be enhanced by spin trapping. Overall, EPR is a versatile methodology for studying NO metabolism in plants.

2.1
Introduction

2.1.1
Chemistry of Nitrogen-Active Species

Nitric oxide (NO) is an inorganic free radical gaseous molecule that has been shown in recent decades to play an unprecedented range of roles in biological systems. The

Nitric Oxide in Plant Physiology. Edited by S. Hayat, M. Mori, J. Pichtel, and A. Ahmad

ISBN: 978-3-527-32519-1

broader chemistry of NO involves a redox array of species with distinctive properties and reactivities: NO^+ (nitrosonium), NO^- (nitroxyl anion), and $NO^•$ (nitroxyl radical) [1]. Neutral NO has a single electron in its 2p-π antibonding orbital and the removal of this electron forms NO^+, while the addition of one more electron to NO forms NO^- [2].

NO^+ (nitrosonium). The chemistry of NO^+ is characterized by addition and substitution reactions with nucleophiles such as electron-rich bases and aromatic compounds. The NO^+ ion has been described as responsible for the formation of nitroso compounds [2].

NO^- (nitroxyl anion). NO^- converts rapidly to N_2O through dimerization and dehydration [3] and is known to react with Fe(III) heme [4]. NO^- also undergoes reversible addition to both low molecular weight and protein-associated thiols, leading to sulfhydryl oxidation. Electron transfer and collisional detachment reactions are common and generally yield NO radical as the major product. *S*-Nitrosothiols are believed to be a minor product of the reaction of NO^- with disulfides [2].

$NO^•$ (NO radical). From a biological point of view, the important reactions of NO are those with oxygen and its various redox forms and with transition metal ions.

$$NO^• + O_2{}^- \rightarrow ONOO^- \tag{2.1}$$

Reaction (2.1) is a second-order reaction with respect to NO according to Eq. (2.2), at a rate close to diffusion.

$$v = k[NO]^2[O_2{}^-] \tag{2.2}$$

The biological half-life of NO, generally assumed to be in the order of seconds, strongly depends on its initial concentration. In aqueous solution, NO can undergo auto-oxidation (i.e., reaction with O_2) to produce N_2O_3 and this compound can undergo hydrolysis to form nitrite [5]. Since NO and O_2 are 6–20 times more soluble in lipid layers compared to aqueous fractions, the rate of auto-oxidation is expected to increase in the lipid phase [5] and the primary reactions of N_2O_3 are thought to occur primarily in the membrane fraction.

2.1.2
Biological Effects of NO

When discussing the chemistry and physiological effects of NO, it should be appreciated that NO is a highly diffusible second messenger that can elicit effects relatively far from its site of production. The concentration and therefore the source of NO are the major factors determining its physiological effects [6]. At low concentrations ($<1\,\mu M$), direct effects of NO predominate. At higher concentrations ($>1\,\mu M$), indirect effects, mediated by reactive nitrogen species (RNS), prevail.

Direct effects of NO most often involve the interaction of NO with metal complexes. NO forms complexes with transition metal ions, including those regularly found in metalloproteins. Reactions with heme-containing proteins have been widely studied, particularly in the case of hemoglobin. NO also forms nonheme transition metal complexes and interest has been focused on its reactions with iron–sulfur centers in proteins including several involved in mitochondrial electron transport and enzymes such as aconitase [7]. The reactions of NO with heme-containing proteins are physiologically the most relevant and include interactions with soluble guanylate cyclase [8] and cytochrome P450 [9]. NO is also able to terminate lipid peroxidation [10].

Indirect effects of NO, produced through the interaction of NO with either O_2 or O_2^-, include nitrosation (when NO^+ is added to an amine, thiol, or hydroxy aromatic group), oxidation (when one or two electrons are removed from the substrate), or nitration (when NO_2^+ is added to a molecule) [6]. $ONOO^-$ acts as both nitrating agent and powerful oxidant to modify proteins (oxidation and formation of nitrotyrosine), lipids (lipid oxidation and lipid nitration), and nucleic acids (DNA oxidation and DNA nitration).

2.2 Methods of NO Detection

Perhaps, one of the most productive and exciting fields of biochemistry and physiology since its discovery 20 years ago is the study of NO [11]. Several techniques have been developed to assess NO concentration by amperometric, chemiluminescent, fluorescent, and spectrophotometric methods and by electron paramagnetic resonance (EPR) detection [12].

2.2.1 Determination of NO by Specific Electrodes

The oxygen monitor method is an economical and accurate approach to quantify NO concentration in stock solutions within minutes [13]. This method is based on the reaction of NO with O_2 in liquid phase (reaction (2.3)), measuring the consumption of O_2 by NO. Even though the limit of detection is approximately 10 μM to 1.9 mM in solution, this method was used in studies on the biochemistry of NO [13]. For electrochemical detection of NO, a selective electrode in conjunction with a DUO 18 data acquisition system (ISO-NO Mark II WPI, USA) is employed [14].

$$4\,NO + 2\,H_2O + O_2 \rightarrow 4\,HNO_2 \qquad (2.3)$$

2.2.2 Determination of NO by Spectrophotometric and Fluorometric Methods

It is known that NO is relatively unstable in the presence of molecular oxygen and will rapidly and spontaneously auto-oxidize to yield a variety of nitrogen oxides (NO_2, N_2O_3, and NO_2^-). Moreover, it is suggested that the NO_2^- derived from NO auto-oxidation is rapidly converted to NO_3^-. One method for the indirect determination of NO involves the spectrophotometric measurements of its stable decomposition products NO_2^- and NO_3^- by the Griess reaction [12]. This is a two-step diazotization reaction in which the NO-derived nitrosating agent, generated from the acid-catalyzed formation of nitrous acid and nitrite, reacts with sulfanilic acid to produce a diazonium ion. Diazonium is then coupled to *N*-(1-napthyl)ethylene diamine to form chromophoric azo compounds that absorb strongly at 543 nm [15]. Quantification of NO_2^- and NO_3^- was described via the enzymatic reduction of NO_3^- to NO_2^- using a commercially available nitrate reductase (NR). The Griess reaction is a simple, rapid, and inexpensive assay; however, its practical sensitivity limit is only on the order of 2–3 μM. Since NO_3^- occurs in high concentrations such as 20 mM in vacuoles and 1–5 mM in the cytosol of plant tissue [16], the Griess reaction has severe limitations for use in plants. To enhance the sensitivity of measuring NO_2^-, the aromatic diamino compound 2,3-diaminonaphthalene (DAN) was used [17]. This fluorometric assay offers the additional advantages of specificity, sensitivity, and versatility; moreover, the assay is capable of detecting levels as low as 10–30 nM. In addition to the DAN assay, it has been demonstrated that diaminofluoroscein-2 diacetate (DAF-2DA) may be used to determine the presence of NO *in vitro* and *in situ* [18].

2.2.3
Determination of NO by Electron Paramagnetic Resonance

Electron paramagnetic resonance is considered the most powerful technique for the detection and identification of biological radicals [11, 19]. Like oxygen, NO is paramagnetic, with the free electron shared between nitrogen and oxygen. Although EPR technique can report on levels of NO *in vitro* and *in vivo*, it is not regarded as the simplest and most practical means of measuring NO (given that other techniques can be used for multiple sampling, often requiring no specialized equipment); it is, however, certainly the only method by which NO and its paramagnetic derivatives can be unambiguously identified with a low detection limit ($\sim 10^{-9}$ M) without interference, as observed when other methods are applied [13]. DETC (diethyldithiocarbamate), MGD (*N*-methyl-D-glucamine dithiocarbamate), and DTCS (dithiocarboxy sarcosine) were used as trapping reagents [20].

An accurate technique for NO detection is critical for a complete understanding of the physiological and pathophysiological processes in which NO is implicated.

2.2.3.1 Specific Experimental Advances

A study was undertaken in our laboratory to formulate an assay for assessing endogenous NO generation in actively developing tissue such as soybean and sorghum embryonic axes [21, 22]. Experimental conditions were carefully selected. To optimize NO determination, spin trapping was employed to detect NO.

An Fe^{2+} complex with MGD (Fe^{2+}-MGD_2) acts as a high-affinity NO spin trap agent (Figure 2.1a). A distinctive and unique EPR signal ($g = 2.03$ and $a_N = 12.5$ G) was obtained when NO was generated from GSNO (0.1 mM) and when Fe-MGD

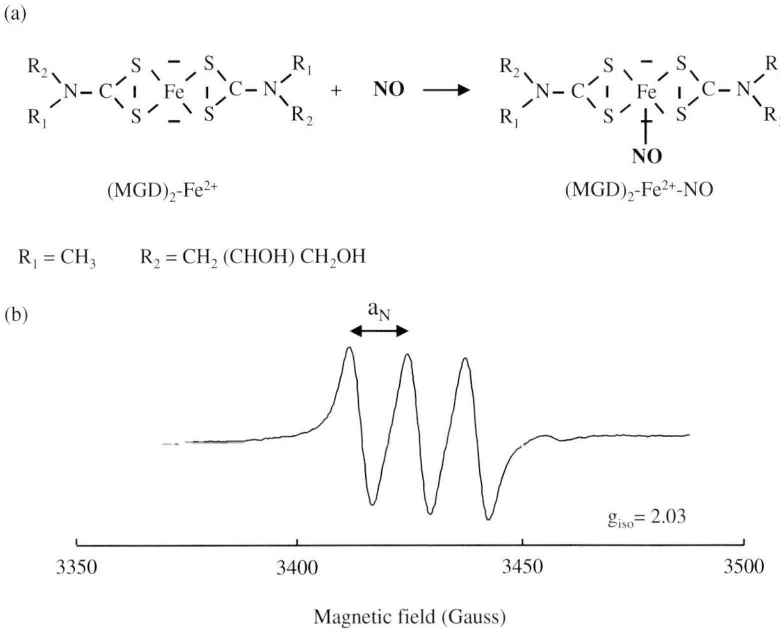

Figure 2.1 EPR spin trapping detection of NO. (a) Chemical reaction between $(MGD)_2$-Fe^{2+} complex and NO. (b) Typical three-line EPR spectrum corresponding to $(MGD)_2$-Fe^{2+}-NO adduct obtained by mixing *S*-nitrosoglutathione 0.1 mM, MGD 5 mM, and $FeSO_4$ 0.5 mM at room temperature.

(0.5–5 mM) was supplemented at room temperature (Figure 2.1b). The reaction was performed in 100 mM phosphate buffer containing the spin trap (5 mM MGD, 0.5 mM $FeSO_4$) [23]. Samples were transferred to bottom-sealed Pasteur pipettes prior to EPR analysis. The spectra were recorded at room temperature (18 °C) in a Bruker ECS 106 EPR spectrometer operating at 9.5 GHz. Instrument settings included 150 G field scan, 83.886 s sweep time, 327.68 ms time constant, 5.983 G modulation amplitude, 50 kHz modulation frequency, and 20 mW microwave power. To select these parameters, effects of the modulation amplitude (Figure 2.2a) and microwave power were tested (Figure 2.2b).

The distinctive and unique EPR signal ($g = 2.03$ and $a_N = 12.5$ G) that enables a fingerprint-like identification of NO was detected in the homogenates from both soybean embryonic axes and cotyledons, and sorghum embryonic axes (Figure 2.3). Quantification of the spin adduct was performed using an aqueous solution of 4-hydroxy-2,2,6,6-tetramethyl piperidine-1-oxyl (TEMPOL). TEMPOL is a stable free radical used as a standard to obtain the concentration of other free radical adducts. TEMPOL solutions were standardized spectrophotometrically at 429 nm using $\varepsilon = 13.4\,M^{-1}\,cm^{-1}$. Data in Table 2.1 show the steady-state concentration of NO, estimated as the concentration of $(MGD)_2$-Fe^{2+}-NO adduct obtained by double integration of the three-line spectra and cross-checked with the TEMPOL spectra.

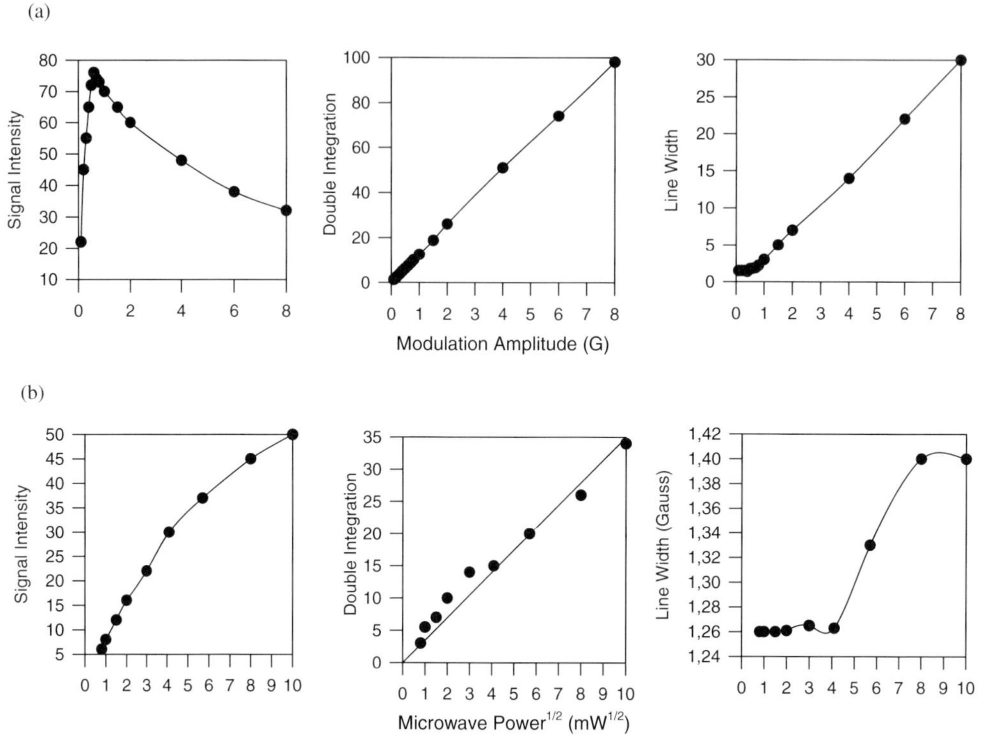

Figure 2.2 Effect of instrumental parameters on NO EPR detection. (a) Influence of modulation amplitude on $(MGD)_2$-Fe^{2+}-NO complex signal properties: signal intensity, double integration, and line width. (b) Effect of microwave power on $(MGD)_2$-Fe^{2+}-NO complex EPR signal properties.

2.3 Use of EPR Methodology for Assaying Enzyme Activities

In animals, it is widely accepted that NO production is predominantly catalyzed by the activity of nitric oxide synthases (NOSs), which are heme-containing proteins related to the cytochrome P450 family. These enzymes catalyze the conversion of L-arginine (Arg) to L-citrulline and NO, using NADPH and O_2 as cosubstrates and employing FAD, FMN, tetrahydrobiopterin (BH_4), and calmodulin (CaM) as cofactors [24, 25]. Cueto *et al.* [26] and Ninnemann and Maier [27] were the first to show the activity of NOS-like enzymes in leguminous plants. Since then, a significant number of reports have shown the presence of NOS-like enzymes in different plant species. Recently, an NOS-like activity was reported in pea stems referred to as N^{ω}-nitro-L-arginine (LNNA)-sensitive NO generation by a hemoglobin assay [28]. The activities reported were about 15 and 4 nmol min^{-1} mg^{-1} protein for microsomal and cytosolic fractions, respectively. Electron microscopy immunogold labeling with antibodies made

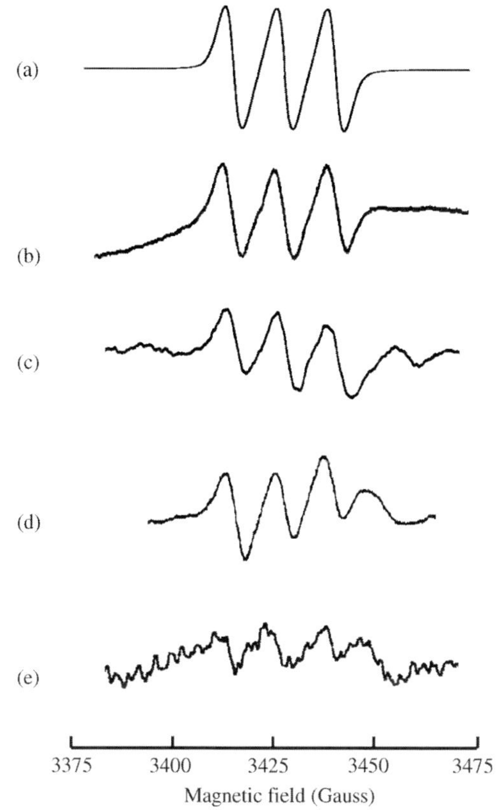

Figure 2.3 Typical electron paramagnetic resonance spectra of the $(MGD)_2$-Fe^{2+}-NO spin adduct. (a) Computer-simulated EPR spectrum using the parameters $g = 2.03$ and $a_N = 12.5$ G. (b) Typical EPR spectrum of Antarctic *Chlorella* sp. cells registered after 6 days of growth. (c) Typical EPR spectrum of sorghum (*Sorghum bicolor*) embryo excised from seeds after 30 h of imbibition in a medium containing 12 mM KNO_3. (d) Typical EPR spectrum of soybean (*Glycine max*) embryonic axes excised from seeds after 24 h of imbibition in a medium containing 15 mM KNO_3. (e) Typical EPR spectrum of soybean (*Glycine max*) cotyledons from 6-day-old seedlings.

against rabbit brain NOS or murine iNOS has confirmed the subcellular localization of an NOS-like protein in the matrix of peroxisomes, in chloroplasts, and in the nucleus of tobacco, maize, and pea [29, 30]. The presence of NO was also confirmed employing the fluorescence dye DAF-2DA and by ozone chemiluminescence [31]. This NOS-like activity is apparently immunorelated to animal iNOS and the activity in peroxisomes was about 5 nmol NO min^{-1} mg^{-1} protein. Tobacco cells treated with cryptogein showed early increases in NO-dependent DAF-2DA fluorescence located in chloroplasts occurring within the first 3 min [32]. More recently, by assessing the presence of peroxisomal NO by EPR [31], the activity of an NOS-like enzyme was suggested.

Table 2.1 EPR determination of NO content and arginine or nitrite-dependent NO production in different plant organs and species.

	NO steady-state concentration	Maximum NO generation rate		Reference
		Arg dependent	NO_2^- dependent	
Sorghum embryos (24 h)	8.0 ± 0.8 (nmol g^{-1} FW)	2.2 ± 0.3 (nmol min^{-1} g^{-1} FW)	18 ± 3 (nmol min^{-1} g^{-1} FW)	[21]
Soybean embryonic axes (15 h)	0.38 ± 0.05 (nmol g^{-1} FW)	—	—	[20]
Soybean cotyledons (6 d)	6 ± 1 (nmol g^{-1} FW)	—	1.4 ± 0.5 (nmol min^{-1} g^{-1} FW)	[54]
Antarctic *Chlorella* sp. (6 d)	156 ± 31 (pmol 10^7 $cell^{-1}$)	4.0 ± 0.7 (pmol min^{-1} 10^7 $cell^{-1}$)	9 ± 1 (pmol min^{-1} 10^7 $cell^{-1}$)	[50]
Soybean leaves chloroplasts	nd	0.76 ± 0.04 (nmol min^{-1} mg^{-1} protein)	3.2 ± 0.2 (nmol min^{-1} mg^{-1} protein)	[14]

The following plant material was employed: sorghum (*Sorghum bicolor*) embryonic axes from seeds imbibed for 24 h, soybean (*Glycine max*) embryonic axes from seeds imbibed for 15 h, cotyledons from soybean seedlings developed for 6 days, Antarctic *Chlorella* sp. cultures developed for 6 days, and chloroplasts isolated from 10-day-old soybean seedlings. For NO content, the tissues were homogenized in the presence of the complex $(MGD)_2$-Fe^{2+}. For NO generation, the samples were added with the spin trap and the corresponding substrates and cofactor, and then incubated up to 10 min. nd: nondetectable.

2.3.1 NOS-Like Dependent NO Generation

To assess the possible contribution of endogenous NO synthase activity to the generation of NO, an EPR assay was developed [22]. Sorghum embryonic axes homogenates were supplemented with $(MGD)_2$-Fe^{2+}, NADPH, and L-arginine, and the online rate of NO generation was measured as the increase in spectrum peak height at 3412 G for 10 min (Figure 2.4a). Under these conditions, the online rate of generation of the adduct $(MGD)_2$-Fe^{2+}-NO was 2.2 ± 0.3 nmol min^{-1} g^{-1} FW.

2.3.2 Nitrate Reductase-Dependent NO Generation

Cytosolic nitrate reductase was the first enzymatic source of NO to be identified. Assimilatory NR is a large homodimeric enzyme with a molecular mass between 200 and 250 kDa. Each monomer houses three prosthetic groups, FAD, heme, and molybdenum cofactor, which transfer electrons from NADH to the enzyme substrate (normally NO_3^-). NR is structurally similar to NOS from animal tissues since it

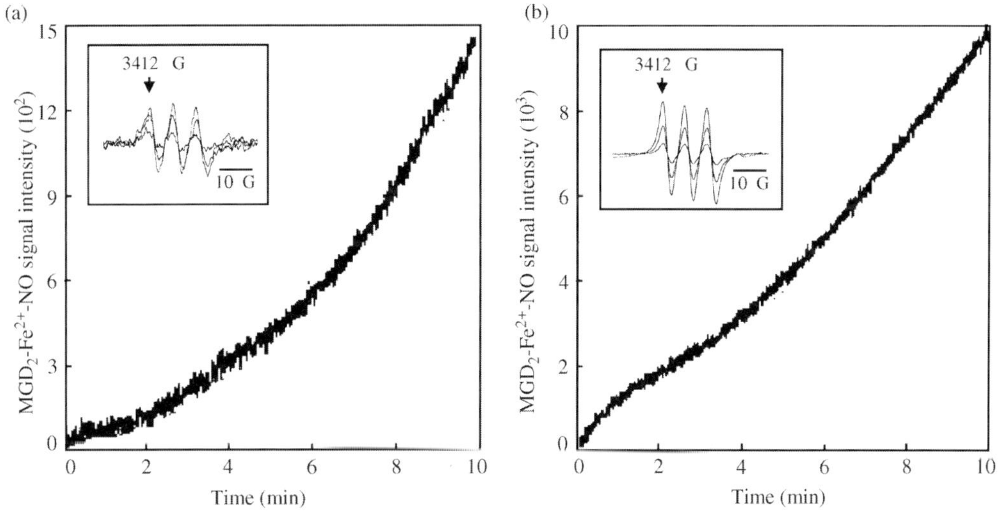

Figure 2.4 EPR detection of enzymatic NO generation. (a) Online $(MGD)_2$-Fe^{2+}-NO spin adduct generation by NOS-like activity in sorghum embryonic axes. The homogenates were supplemented with 1 mM Arg, 5 mM $MgCl_2$, 1 mM $CaCl_2$, and 0.1 mM NADPH. The height of the peak at 3412 G was recorded during 10 min. Inset: spectra measured at 4, 8, and 10 min. (b) Online $(MGD)_2$-Fe^{2+}-NO spin adduct generation by NR activity in sorghum embryonic axes. The homogenates were supplemented with 1 mM NO_2^- and 0.1 mM NADH. The height of the peak at 3412 G was recorded during 10 min. Inset: spectra measured at 6, 8, and 10 min.

contains a heme and an NAD(P)H/FAD binding domain. It should be noted that both NR and NOS are flavoenzymes and members of the ferredoxin-$NADP^+$ reductase family [33]. It has been shown that NR, apart from its role in NO_3^- reduction and assimilation, can also catalyze the reduction of NO_2^- to NO. It was demonstrated that *in vivo* NR activity is responsible for releasing NO [34, 35]. These experiments were conducted under anaerobic conditions and resulted in NO_2^- accumulation, and it was established that NR was needed for NO formation [36, 37]. It was also demonstrated that cytosolic NADH-dependent NR from maize is capable of reducing not only NO_3^- to NO_2^- but also NO_2^- to NO, which opened up the possibility for cytosolic NR to be involved in NO production in plants [38, 39]. This line of experiments culminated in the demonstration that purified NR reduces NO_2^- to NO [38–40]. However, for maize NR, the K_m for NO_2^- is between 100 and 300 μM, whereas NO_3^- is a competitive inhibitor with a K_i of 50 μM [40]. Thus, under physiological conditions where NO_3^- levels are high (1–5 mM) and NO_2^- content is low (10 μM in illuminated leaves), a significant NO production from NR would not be expected [41]. On the other hand, under anaerobic conditions, NO_2^- levels increase and NO release can increase 100-fold [40]. In this regard, high NO levels can be generated using nitrite reductase (NiR)-deficient lines that accumulate NO_2^- [42, 43].

Simontacchi *et al.* [22], using the same experimental protocol employed to assess NO generation by the NOS-like activity, identified NO by EPR as a product from the activity of the enzyme NR in sorghum axes. Homogenates from axes isolated from sorghum seeds incubated for 24 h in the presence of 12 mM nitrate were supplemented with $(MGD)_2$-Fe^{2+}, NADH, and the substrates for NADH-dependent NR activity, and the rate of generation of NO was measured by EPR to assess maximal online NO generation by enzymatic activity. The $(MGD)_2$-Fe^{2+}-NO spectrum was recorded for 10 min and the increase in peak height at 3412 G was plotted (Figure 2.4b). In the presence of 1 mM NO_2^-, the rate of NO generation was 18 ± 3 nmol $min^{-1} g^{-1}$ FW. The availability of NO_3^- (from 4 to 30 mM) during imbibition of sorghum seeds strongly affected the NR-dependent rate of generation of NO in the embryonic axes [22].

2.4
Application of EPR Methods to Assess NO Generation During Plant Development

To fully understand the complexity of NO biological effects, the first aspect to be considered is that plant tissues are in contact with both external (atmospheric NO and soil NO) and internally generated NO (endogenous NO).

During evolution, plants developed a special organ, the seed, which ensures their dispersion. Germination, usually defined as the rupture of the seed coat by the radicle and/or coleoptile, increases the rate of NO_2^- entry from the soil solution, and this source of NO_2^- may be augmented by NO_2^- produced enzymatically by NR [44]. Either NO_2^- or sodium nitroprusside (SNP) improved the germination rate of *Suaeda salsa* under saline conditions [45]. Moreover, nitrogenous compounds such as organic and inorganic NO_3^- promote germination in many seeds [46], and it has been shown that NO_3^- significantly increases light sensitivity and decreases light requirement for seeds [47]. Germination of photoblastic lettuce seeds is a phytochrome-dependent process above 25 °C, and it was demonstrated that NO donors such as SNP are able to promote germination in the dark to the same extent as either gibberellin treatment or a 5-min pulse of white light [48]. However, seeds were also able to germinate in the light, in the presence of the NO scavenger cPTIO, suggesting that light and NO can stimulate germination via different pathways [48]. On the other hand, NO donors enhance the phytochrome A-dependent germination response to red light of photodormant *Arabidopsis* seeds [49].

In this scenario, the EPR techniques shown here were applied to assess NO steady-state concentration in sorghum and soybean embryonic axes during the initial stages of germination (Table 2.1). NO content, at its maximum value over the imbibition time, was 22-fold higher in sorghum compared to soybean embryonic axes (Table 2.1). Other photosynthetic species, such as Antarctic *Chlorella* sp. and Antarctic *Chlamydomonas* sp., showed a maximum NO content before the beginning of the lag phase of growth of the cultures (Table 2.1); however, *Chlorella vulgaris* sp. did not show this behavior [50]. This finding suggested that NO could be part of the signaling network operative under specific growth conditions, such as low temperature stress, and could be an adaptive mechanism to allow optimal cell survival. Data in Table 2.1 also show the maximum generation rate, assessed by EPR, dependent on either arginine or NO_2^- supplementation. In all the tested systems, both activities are within the same range; however, NO_2^--dependent activity seems to be higher than arginine-dependent enzymatic activity (Table 2.1). Moreover, NO production has previously been detected in several compartments within the plant cell [51]. By EPR spectroscopy, evidence has demonstrated that chloroplasts participate in NO synthesis in plants as was suggested by previous nonquantitative studies employing fluorescence microscopy [32, 52] and immunogold electron microscopy [29].

2.5 Conclusions

Detection of NO is a challenge, in particular because of the difficult dual requirements for specificity and sensitivity. EPR spectroscopy has been one of the first methods used before 1990 for NO detection in response to requirements for greater specificity [53]. The $(MGD)_2$-Fe^{2+} complex that allowed the first real-time detection of NO in a live animal over long periods (several hours) [23] also could be successfully used to detect NO in plant systems, as shown in this chapter. However, it should be pointed out that careful experimental controls and alternative methods should be included to arrive at sound and consistent conclusions. The main advantages of EPR methods, such as good potential specificity and reasonable time resolution, overcome the poor sensitivity that could be enhanced by spin trapping. Overall, EPR is a versatile methodology to use for studying NO metabolism in plants.

References

1 Gisone, P., Dubner, D., Perez, M.R., Michelin, S., and Puntarulo, S. (2004) *In Vivo*, **18**, 281–292.
2 Stamler, J.S., Singel, D.J., and Loscalzo, J. (1992) *Science*, **258**, 1898–1902.
3 Basylinski, D.A. and Hollocher, T.C. (1985) *Inorg. Chem.*, **24**, 4285–4288.
4 Goretski, J. and Hollocher, T.C. (1988) *J. Biol. Chem.*, **263**, 2316–2323.
5 Ford, P.C., Wink, D.A., and Sanbury, D.M. (1993) *FEBS Lett.*, **326**, 1–3.
6 Wink, D.A. and Mitchell, J.B. (1998) *Free Radic. Biol. Med.*, **25**, 434–456.
7 Henry, Y., Durocq, C., Darpier, J.C., Servent, D., Pellac, C., and Guissani, A. (1991) *Eur. Biophys.*, **20**, 1–15.
8 Martin, E., Davis, K., Bian, K., Lee, Y.C., and Murad, F. (2000) *Semin. Perinatol.*, **24**, 2–6.
9 Wink, D.A., Osawa, Y., Darbyshire, J.F., Jones, C.R., Eshenaur, S.C., and Nims, N.W. (1993) *Arch. Biochem. Biophys.*, **300**, 115–123.
10 Rubbo, H., Parthasarathy, S., Barnes, S., Kirk, M., Kalyanaraman, B., and Freeman, B.A. (1995) *Arch. Biochem. Biophys.*, **324**, 15–25.
11 Jackson, S.C., Hancock, J.T., and James, P.E. (2007) *Electron Paramagn. Reson.*, **20**, 192–244.
12 Tarpey, M.M., Wink, D.A., and Grisham, M.B. (2004) *Am. J. Physiol. Regul. Integr. Comp. Physiol.*, **286**, R431–R444.
13 Venkataraman, S., Martin, S.M., Schafer, F.Q., and Buettner, G.R. (2000) *Free Radic. Biol. Med.*, **29**, 580–585.
14 Jasid, S., Simontacchi, M., Bartoli, C.G., and Puntarulo, S. (2006) *Plant Physiol.*, **142**, 1246–1255.
15 Grisham, M.B., Johnson, G.G., and Lancaster, J.R., Jr. (1996) *Methods Enzymol.*, **268**, 237–246.

16 Crawford, N.M., Kahn, M.L., Leustek, T., and Long, S.R. (2000) Nitrate reduction, in *Biochemistry and Molecular Biology of Plants* (eds B.B. Buchanan, W. Gruissem, and R.L. Jones), American Society of Plant Physiologists, Rockville, MD, pp. 818–821.

17 Miles, A.M., Wink, D.A., Cook, J.C., and Grisham, M.B. (1996) *Methods Enzymol.*, **268**, 105–120.

18 Kojima, H., Nakatsubo, N., Kikuchi, K., Kawahara, S., Kirino, Y., Nagoshi, H., Hirata, Y., and Nagano, T. (2001) *Anal. Chem.*, **73**, 1967–1973.

19 Becker, C.F., Lausecker, K., Balog, M., Kálai, T., Hideg, K., Steinhoff, H.J., and Engelhard, M. (2005) *Magn. Reson. Chem.*, **43**, S34–S39.

20 Xu, Y.C., Cao, Y.L., Guo, P., Tao, Y., and Zhao, B.L. (2004) *Phytopathology*, **94**, 402–407.

21 Caro, A. and Puntarulo, S. (1999) *Free Radic. Res.*, **31**, S205–S212.

22 Simontacchi, M., Jasid, S., and Puntarulo, S. (2004) *Plant Sci.*, **167**, 839–847.

23 Komarov, A.M. and Lai, C.S. (1995) *Biochim. Biophys. Acta*, **1272**, 29–36.

24 Stuehr, D.J. (1999) *Biochim. Biophys. Acta*, **1411**, 217–230.

25 Bogdan, C. (2001) *Trends Cell Biol.*, **11**, 66–75.

26 Cueto, M., Hernández-Perera, O., Martín, R., Bentura, M.L., Rodrigo, J., Lamas, S., and Golvano, M.P. (1996) *FEBS Lett.*, **398**, 159–164.

27 Ninnemann, H. and Maier, J. (1996) *Photochem. Photobiol.*, **64**, 393–398.

28 Qu, Y., Feng, H., Wang, Y., Zhang, M., Cheng, J., Wang, X., and An, L. (2006) *Plant Sci.*, **170**, 994–1000.

29 Barroso, J.B., Corpas, F.J., Sandalio, L.M., Valderrama, R., Palma, J.M., Lupiáñez, J.A., and del Río, L.A. (1999) *J. Biol. Chem.*, **274**, 36729–36733.

30 Ribeiro, E.A., Cunha, F.Q., Tamashiro, W.M.S.C., and Martins, I.S. (1999) *FEBS Lett.*, **445**, 283–286.

31 Corpas, F.J., Barroso, J.B., Carreras, A., Quiros, M., Leon, A.M., Romero-Puertas, M.C., Esteban, F.J., Valderrama, R., Palma, J.M., Sandalio, L.M., Gomez, M., and del Rio, L.A. (2004) *Plant Physiol.*, **136**, 2722–2733.

32 Foissner, I., Wendehenne, D., Langebartels, C., and Durner, J. (2000) *Plant J.*, **23**, 817–824.

33 Lamattina, L., Garcia-Mata, C., Graziano, M., and Pagnussat, G. (2003) *Annu. Rev. Plant Biol.*, **54**, 109–136.

34 Harper, J.E. (1981) *Plant Physiol.*, **68**, 1488–1493.

35 Klepper, L.A. (1987) *Plant Physiol.*, **85**, 96–99.

36 Dean, J.V. and Harper, J.E. (1988) *Plant Physiol.*, **88**, 389–395.

37 Klepper, L.A. (1990) *Plant Physiol.*, **93**, 26–32.

38 Yamasaki, H., Sakihama, Y., and Takahashi, S. (1999) *Trends Plant Sci.*, **4**, 128–129.

39 Yamasaki, H. and Sakihama, Y. (2000) *FEBS Lett.*, **468**, 89–92.

40 Rockel, P., Strube, F., Rockel, A., Wildt, J., and Kaiser, W.M. (2002) *J. Exp. Bot.*, **53**, 103–110.

41 Crawford, N.M. (2006) *J. Exp. Bot.*, **57**, 471–478.

42 Morot-Gaudry-Talarmain, Y., Rockel, P., Moureaux, T., Quilleré, I., Leydecker, M.T., Kaiser, W.M., and Morot-Gaudry, J.F. (2002) *Planta*, **215**, 708–715.

43 Planchet, E., Gupta, K.J., Sonoda, M., and Kaiser, W.M. (2005) *Plant J.*, **41**, 732–743.

44 Ferrari, T.E. and Varner, J.E. (1970) *Proc. Natl. Acad. Sci. USA*, **65**, 729–736.

45 Li, W., Liu, X., Khan, M.A., and Yamaguchi, S. (2005) *J. Plant Res.*, **118**, 207–214.

46 Grubišić, D., Giba, Z., and Conjević, R. (1992) *Photochem. Photobiol.*, **56**, 629–632.

47 Toole, E.H., Toole, V.K., Bortwick, H.A., and Hendricks, S.B. (1955) *Plant Physiol.*, **30**, 15–21.

48 Beligni, M.V. and Lamattina, L. (2000) *Planta*, **210**, 215–221.

49 Batak, I., Dević, M., Giba, Z., Grubišić, D., Poff, K., and Conjević, R. (2002) *Seed Sci. Res.*, **12**, 253–259.

50 Estevez, M.S. and Puntarulo, S. (2005) *Phys. Plant*, **125**, 192–201.

51 Lamotte, O., Courtois, C., Barnavon, L., Pugin, A., and Wendehenne, D. (2005) *Planta*, **221**, 1–4.

52 Gould, K.S., Lamotte, O., Klinguer, A., Pugin, A., and Wendehenne, D. (2003) *Plant Cell Environ.*, **26**, 1851–1862.

53 Henry, Y. and Guissani, A. (2000) *Analusis*, **28**, 445–454.

54 Simontacchi, M., Galatro, A., Jasid, S., and Puntarulo, S. (2006) *Curr. Top. Plant Biol.*, **7**, 19–25.

3
Calcium, NO, and cGMP Signaling in Plant Cell Polarity

Ana Margarida Prado, José A. Feijó, and David Marshall Porterfield

Summary

In addition to calcium, nitric oxide and cGMP are two of the most ubiquitous signal molecules, playing a major role in controlling nearly every cellular and organ function in mammals. The link between calcium, nitric oxide, and cGMP signaling is well understood in animals, and now is emerging as a major component in two model plant gametophyte systems, pollen tubes and fern spores. Localized calcium gradients are necessary during the process of developmental polarity in both of these haploid, single-cell systems. While research has shown that NO/cGMP signaling cascades are active components of cellular polarity and tip growth in these systems, the detailed connection between calcium and NO/cGMP is emerging only now. In the animal model, calcium precedes NO/cGMP and a calcium–calmodulin complex bridges the two separate redox centers in the animal NOS enzyme. A significant missing piece of the plant model is the NO synthesizing enzyme, which could have a calcium-dependent regulatory component, if the animal model is also found to be active in plants. Still, there are other probable models, including a mechanism where NO directly modulates the cytoskeleton.

3.1
Introduction

Nitric oxide (NO), a ubiquitous signaling molecule in animals, has been implicated in numerous plant signaling systems. With regard to a central and fundamental role of NO in plant cell development, recent work has focused on understanding the role of NO and cGMP in two model plant gametophyte systems, that is, pollen tubes and fern spores. Both of these haploid, single-cell systems are known to developmentally depend on localized calcium gradients that actively drive cellular polarity. Although

Nitric Oxide in Plant Physiology. Edited by S. Hayat, M. Mori, J. Pichtel, and A. Ahmad

ISBN: 978-3-527-32519-1

the research that has been conducted in this area has demonstrated that NO/cGMP signaling cascades are active components in calcium-dependent signaling associated with cellular polarity and tip growth in these systems, the detailed connection between calcium and NO/cGMP is being addressed in earnest only now.

Why is the connection between calcium and NO in plant systems important? The answer to this question is in the model of NO signaling that is well documented and described in animal systems (Figure 3.1). In this model, NO and cGMP cascades are actually preceded and controlled by calcium via activated calmodulin. The calcium–calmodulin complex bridges the two separate redox centers in the animal NOS enzyme, thereby regulating the conversion of arginine and oxygen to citruline and NO. One major roadblock that the plant side of the NO world has faced is the fact that a definitive discovery of the plant source for NO production has not yet been clearly made. In fact, the early claims of identification and cloning of the plant NOS have resulted in a series of high-profile papers followed by high-profile retractions. Instead of focusing on the source of plant NO synthesis, this chapter is intended to delineate the overall signaling pathway and attempt to define how this involves calcium and how it is a central theme in plant development and reproduction.

Why investigate NO as an intracellular signaling component in these plant systems? NO is a small and one of the simplest biologically active molecules in nature, and one of the simplest signaling systems, as it is self-regulating via an auto-oxidation mechanism. Besides calcium, nitric oxide also appears to be one of the most

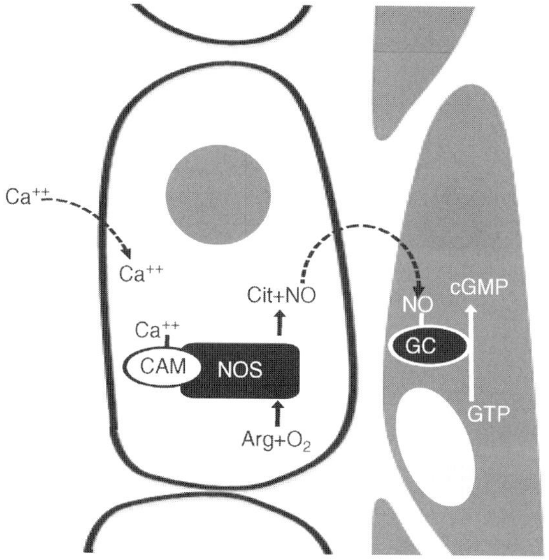

Figure 3.1 (a) Overview of basic calcium-dependent NO/cGMP signaling in mammalian vascular endothelium. A ligand-gated calcium channel opens and the cytoplasmic free calcium increases, activating calmodulin. The calcium–calmodulin complex then activates eNOS, producing NO that in turn activates guanlyate cyclase activity. Calcium is opposed by calcium pumps, NO is opposed by auto-oxidation, and cGMP is opposed by phosphodiesterase activity.

ubiquitous substances, playing a major role in controlling nearly every cellular and organ function in mammals [1]. While NO is clearly a fundamental signaling molecule in animal biology, it is now emerging as a major signaling component in many plant physiological processes (reviewed in Refs [2–4]). After the complete signaling pathway (Ca^{2+}–NO–cGMP) was revealed in the cardiovascular system in mammals (Figure 3.1), many follow-up discoveries on NO/cGMP signaling were made simply by following calcium through the previous literature.

But what is the connection between calcium and NO in plant systems? In animals, there are numerous examples of calcium-independent NO signaling pathways that emerged after the endothelial NO system was discovered. In plants, the story is turning out to be similar. In some cases, increases in cytosolic calcium precede and activate endogenous NO production, presumably via calmodulin [5–7]. And yet in other cases, NO seems to precede the release of Ca^{2+} from internal stores [7, 8]. Although our understanding of these systems continues to advance, there is still a significant precedence from other eukaryotic systems to believe that calcium may have an involvement in NO/cGMP signaling.

Other reasons for considering NO as a potential signaling molecule in plants is because of the basic characteristics derived from studies in animals (reviewed in Refs [1, 9]): (1) NO diffuses freely across cell membranes, (2) it is known to act as an intra- and intercellular messenger in a number of regulation mechanisms; (3) it is known to act as a positional cue diffusing from point sources; and (4) because it is a gas, it acts at minimal thresholds over considerable distances.

For almost a decade, research in plant biology has expanded the list of examples supporting a central role for NO as a regulatory signaling molecule in plant development and biotic/abiotic stress responses [10, 11]. While, at present, there is no clearly defined and identified plant analogue to the animal NOS enzyme, the universe of NO sources in plant cells is much more versatile than in animal cells. It includes both enzymatic as well as nonenzymatic mechanisms, and the latter can be differentially regulated by the activity of distinct enzymes [11]. In plants, NO has been proposed as a regulator of growth and development processes [10] as exemplified in roots, where NO mediates the response to indole acetic acid during adventitious root formation [12], in senescence by reducing ethylene emission [13], and through the stimulation of seed germination [14]. NO also promotes adaptive responses against drought stress operating downstream from ABA [15], and it has been implicated in the establishment of legume *Rhyzobium* symbiosis [16]. NO plays a role in plant disease resistance by enhancing the induction of hypersensitive response [17–19].

3.2
Cell Polarity and Plant Gametophyte Development

It is difficult to think of an example in eukaryotic systems where cells are not polarized or positioned with respect to other cells in terms of development and tissue integration. There is always an inside–outside, up–down, or right–left directionality in all cells that has to be demarcated early on in cell/tissue development. In complex plant systems, this is initiated in meristematic tissues as patterns of alignment of the cell plate in the latter stages of mitosis and cytokinesis. Tip growth is a dynamic and complex type of cell polarity, which is also common in diverse eukaryotic systems. In fact, fungal mycelium masses, in some cases thought to be the largest and oldest living organisms on the planet, are derived completely by tip-growing hyphae. Moss protonema, root hair, rhizoids in hepatophytes and ferns, pollen tubes, and even developing mammalian neurons all share similar tip growth mechanisms.

In vascular plants, sporophytic meiosis and the alternation of generations have evolved to emphasize the sporophyte phase of the life cycle, and the reduction of the gametophyte, ultimately to just a few cells that depend on the sporophyte for nutrition and protection. Yet, when we compare germination and growth of pollen from an angiosperm to the earliest stages of development of a pterophyte spore, we see that there is much in common in terms of cell polarity and tip growth. It is really not difficult to see how a fern spore could evolve into a pollen grain. Both are the product of meiosis, are resistant and protected *by* a hardened cell wall impregnated with sporopollenin, and undergo polarity development that precedes the emergence of a tip-growing extension. In the fern, this tip-growing cell functions as a rhizoid that anchors the gametophyte and provides water and mineral nutrients. In pollen, this tip-growing extension is the pollen tube, which grows down the floral tissue into the ovary, ultimately to the micropyle of the ovule to deliver sperm nuclei to the egg sac.

3.3
Calcium Signaling in Pollen and Fern Spores

Given the similarities and shared characteristics of pollen and fern spores, it is not surprising that these systems also share conserved mechanisms for signaling and biochemical control. Calcium is very diverse in terms of the vast number of biological processes and systems in which it has been documented. Different modalities of calcium signaling are thought to be programmed through different domains of action that include temporal, spatial, intensity, and frequency.

During germination of spores from the fern *Ceratopteris richardii*, calcium is the earliest measured component of cell polarity. *Ceratopteris* spores can respond to gravity, as an environmental cue to guide polarity and development, ultimately guiding initial rhizoid emergence and growth in alignment with the gravity vector [20]. Calcium signaling associated with this system was first measured using self-referencing calcium microsensors, where a transcellular calcium current, into the bottom and out the top of the spore, was shown to correlate with the timing and magnitude of gravity-guided cell polarization [21]. In other words, the current moves through the cell in alignment with the gravitation vector, like a compass line, to guide cell polarity and the direction of the current. Subsequent studies have shown that reorientation of the cell by inversion follows, almost immediately by reversal of the calcium current [22, 23]. If the calcium current is biochemically inhibited through the use of calcium-specific ion channel inhibitors such as nifedipine, rhizoid orientation is randomized with respect to gravity [21].

A significant role of calcium has also been well documented and explored in both pollen germination and pollen tube growth [24]. While pollen germination, unlike fern spore germination, does not include an obvious early polarity development process, it is clear that pollen requires calcium for germination. Experiments conducted with the vibrating voltage probe measured cation currents correlating with the aperture associated with pollen tube emergence [25]. Calcium was further implicated in experiments using calcium-specific ion channel inhibitors, which were shown to block the germination of pollen [26, 27]. A plasma membrane calcium pump ACA9 [28] and specific calmodulin binding proteins [29] have been shown to be important for pollen germination and seed production.

With regard to pollen tube growth, calcium is a central key player in this dynamic process that can achieve growth rates of up to 4 $\mu m\ s^{-1}$ in some plant species. Associated with the growth of pollen tubes, there is a prerequisite for a tip-localized cytoplasmic calcium gradient, associated with calcium influx at the cell tip [26, 30–40]. Calcium channel blockers have also been shown to be effective inhibitors of pollen tube tip growth, and the tip-localized calcium gradient dissipates when growth stops [41–44].

3.4
NO/cGMP Signaling in Pollen and Fern Spores

We now take it for granted that pollen–pistil communication is driven by a sophisticated signaling system, which to a great extent is unknown [45–47]. Nevertheless, our general understanding of the identity of the specific molecules physiologically active in pollen guidance continues to grow and advance. Nitric oxide signaling in the pollen tube was first documented by Prado *et al.* [48], reporting NO production in pollen. Some of the key experiments were based on a microscale NO gradient in the medium in which the pollen tube was growing. This was accomplished by using a glass micropipette filled with the NO donor *S*-nitroso-*N*-acetylpenicillamine (SNAP) immobilized in agar, thereby allowing NO to diffuse into the medium from a micropipette tip (Figure 3.2). Using this approach, it was demonstrated that externally applied NO acted as a negative chemotropic agent to pollen tube growth [48]. As growing pollen tubes approached the NO gradient near the tip of the micropipette, the growth rate was reduced before reorienting sharply away from the source. This response was blocked by NO scavengers, and it is a significant finding as NO is the only chemotropic agent so far described capable of promoting turning angles of 90° and more (Figure 3.2). This angle is necessary to achieve some steps of the fertilization process, for instance, the entry of the pollen tube in the micropyle in some species (discussed in detail in subsequent sections). Furthermore, enhanced sensitivity to the NO gradient was associated with treatment with sildenafil citrate (Viagra™), an inhibitor of the animal phosphodiesterases, PDE5 and PDE6, suggesting a connection with cGMP signaling in the pollen tube system as well. The importance of NO in fertilization has been recently demonstrated further in *Senecio squalidus* and *Arabidopsis thaliana* [49].

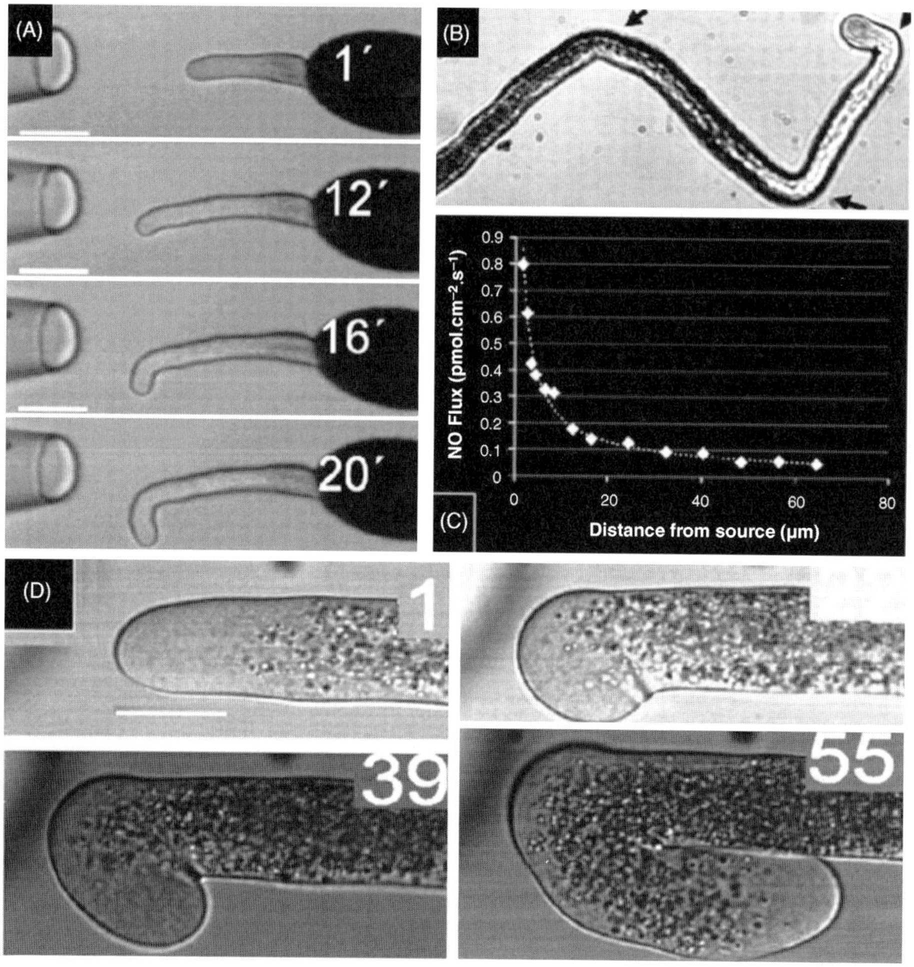

Figure 3.2 (a) Time-lapse sequence of a growing *Lilium longiflorum* (lily) pollen tube facing an extracellular NO point source SNAP on agarose shown on the left of the image). Pollen tube growth slowed as it moved into the NO gradient. When the new growth axis started to be defined, a sharp right angle from the original axis ($97.7° \pm 3.6°$, $n = 28$) was formed. The pollen tube then regained its normal growth rate (bar $= 30\,\mu$m). (b) Lily pollen tube showing three consecutive reorientation responses, which were induced by moving the same source to the locations marked with arrows. (c) Artificial NO source measurements obtained by using a vibrating self-referenced polarographic probe for NO. The graph shows a typical exponential NO-gradient decay from the point source at different step distances. (d) Time-lapse sequence of a pollen tube being challenged with a diluted NO artificial source in the presence of sildenafil citrate (Viagra) (numbers at the upper right-hand corner represent minutes after detection of the response). When these diluted sources are used, most pollen tubes do not show any response, often running into the pipette. For this experiment, pollen tubes were first incubated on standard medium and challenged with the diluted NO source. Despite the lower amount of NO used, reverse reorientation angles were observed in the presence of sildenafil citrate ($109.8° \pm 9.8°$, $n = 9$) showing a sensitization effect from unresponsive to peak response (adapted from Feijo *et al.*, 2004) [50].

More recently, Salmi *et al.* [51] reported that NO is involved in calcium-dependent gravity-directed cell polarity in germinating *C. richardii* fern spores. The approach was based on population statistics and the use of many of the same specific NO-pathway protagonists and antagonists used by Prado *et al.* [48] to study the pollen system. Salmi *et al.* [51] showed that cell polarity in these spores, with respect to gravity, is precariously tuned and disrupted when these signaling pathways are pushed beyond well-defined thresholds. The experimental design included two controls, that is, fixed orientation and a clinostat control, which randomizes the position of the spores relative to the gravity vector by rotating them continuously at 1 rpm. The pharmacology of this system suggests an animal-like NOS activity, as L-NAME, a potent inhibitor of NO-producing nitric oxide synthase, disrupted cell polarity development, and gravity-directed growth of rhizoids. A role downstream of NO for cGMP was also implicated in this system, where the ability of gravity to direct rhizoid growth was completely randomized with LY83583, an inhibitor of mammalian NO activated guanylate cyclase. In addition to disrupting germination polarity and gravity sensing in these cells, the rhizoid morphology was also altered in response to NO/cGMP signaling disruptions (Figure 3.3). Note that the straight and regular growth of the control rhizoids is disrupted, when even low concentrations (1 mM) of these inhibitors are used, and a regular "waving" pattern predominates.

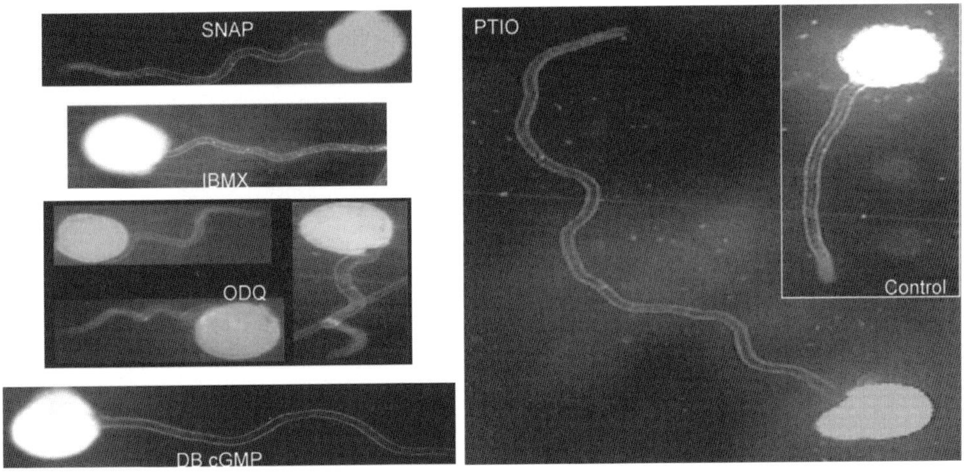

Figure 3.3 *Ceratopteris* rhizoid growth and morphological changes associated with low-concentration exposure (1 μM) to various pharmacological agents for altering NO/cGMP signaling. Note that the control rhizoid is growing downward, exhibiting positive gravity orientation, whereas altering NO/cGMP signaling disrupts gravity orientation. Also, note that the control rhizoid is very straight, whereas the experimental treatments show varying degrees of "rhizoid waving" associated with the pharmacological treatments. In these experiments, SNAP is a NO donor and cPTIO is a cytsolic NO scavenger. Cytosolic cGMP levels were increased directly using DB cGMP or indirectly using the phosphodiesterase inhibitor IBMX. The inhibitor ODQ inhibits guanylate cyclase activity in the rhizoids.

3.5
NO/cGMP in Pollen–Pistil Interactions

Because the pollen tube system has been much more widely studied, it is a better system to further consider the NO/cGMP signaling system. In plants, the pollen tube is a tip-growing cell that evolved to sense subtle extracellular signals and environmental changes and respond to them by changing its growth axis. Early on in pollination, pollen tubes have to communicate their "self" properties (i.e., information about species and individuality) to the stigma cells. This continuous interaction is necessary for success in the process of growth into the open ovary cavity, ultimately to the tiny opening of the micropyle to deliver the sperm nuclei. This is a grand example of the delicacy and intricacy of nature. The functional challenges for the success of the pollen tube in flowering plant reproduction implies an innate ability to communicate with and to decode signals from the floral tissue environment [43, 45]. The investment in the number of transcripts for encoding signaling pathways in pollen tubes verifies and supports the concept of highly coordinated and complex communication processes involved in the process of pollination [43, 45, 52, 53].

For over 140 years, plant biologists have debated the nature of the molecule(s) that presumably guide chemoattraction and navigate the pollen tube toward the ovule, and the debate continues with many open questions [47]. Significant questions on what information is conveyed by the pistil to the pollen, and how does the pollen tube perceive this information, remain to be answered. There is a general agreement that directional growth depends on physical and chemical signals that are exchanged between the male and female gametophytes [54–58]. Yet, despite intense research over the past two decades aimed at defining a mechanistic explanation of the process, consensus around a central or conserved theory of pollen tube guidance is still lacking. Chemotropic mechanisms have been suggested for almost all of the diverse extractable materials from the style (Ca^{2+}, sugars, available water, lipids, arabinogalactan proteins, adhesins, etc.), but none has been sufficiently proven to be viable in this role [59–63] (for review see Refs [45, 57]). The complexities in this system are comparable to those associated with axonal guidance in the nervous system of animals.

The NO pollen guidance hypothesis is supported by recent experiments [64] showing that a dysfunctional peroxin function mutation (*amc*, abstinence by mutual consent) in *Arabidopsis* affects peroxisomal protein import in both pollen and ovules. More importantly, it does not affect pollen tube growth or guidance but prevents pollen tube reception from occurring, thus producing a self-sterile mutant. When both the pollen tube and the ovule carry the *amc* mutation (*amc*/+), pollen reception is substituted by continued growth of the pollen tube that can coil or branch. The *amc*/− ovule can nevertheless attract several pollen tubes. This paper is pivotal because it is the first to show that functional peroxisomes must be present in either the male or female gametophyte for pollen tube guidance to work. This is in agreement with previous work showing that peroxisomes may be the main source of NO production in pollen tubes [48]. The *Atnos1* mutant that was initially thought to be defective in NO production, and to encode for a distinct nitric oxide synthase, was shown to have reduced fertility [65]. More recently, the initial report has been re-examined [66], and as a result, the name of the gene has been revised to NO-associated protein 1 (AtNOA1) because it is now thought to be a member of the circularly permuted GTPase family (cGTPase). When all of these findings are considered together, they support the hypothesis that NO via cGMP is functional as an integral part of the *in vivo* signaling and guidance system.

McInnis *et al.* [49] suggest that NO possibly plays a role as an extracellular signaling molecule, eliciting the reduction of ROS/H_2O_2 in stigmatic papillae. The evidence includes measurement of decreased ROS/H_2O_2 in papillae treated with SNP, a NO donor. This work does not fully support or disapprove a role for a NO-ROS/H_2O_2 signaling system in species recognition or pathogen discrimination between pollen and microorganisms in the stigma.

An important characteristic of pollen tube growth *in vitro*, which supports a role for unknown cues in pollen tube guidance, is the fact that *in vitro* pollen tubes do grow exceedingly straight, without any inherent directionality. It is also worth considering the fact that *in vitro* growth rates are usually significantly slower than those *in vivo* [67]. The TTS proteins are a good example of intervening molecules required for pollen tube growth and guidance. TTS proteins are locally patterned as an apparent path in the style; that is, they do not predict pollen tube directionality [62]. Another such case is the GABA gradient in the *Arabidopsis* pistil, where GABA has been shown to act as a guidance cue *in vivo* whereas it fails to operate in such a manner *in vitro* [68].

The question of whether the genes necessary for NO production are present in the *Arabidopsis* pistil was recently studied. It is necessary to point out that this is a contentious issue as there is no indisputable NO synthase gene known in plants. The most recent candidates include Atnos1/AtNOA1 (Locus At3g47450), which is involved in NO/cGMP signaling, and the nitrate reductases, nr1 (Locus At1g77760) and nr2 (Locus At1g37130), known for the capacity to generate NO from nitrite and contributing to NO-dependent stomatal closure [69]. An analysis of the transcriptomic data gathered by Boavida *et al.* [70] on pollen pistil interactions regarding *Atnos1*, *nr1*, and *nr2* genes revealed differences in transcripts at distinct stages after pollination where both the downregulation of the nr1 transcript and the upregulation of the nr2 transcript occur at 8HAP. This suggests that the differences observed between the expression levels of these transcripts may be correlated with postfertilization events [71].

3.6 Ovule Targeting and NO/cGMP

While there are potential roles for NO signaling in the early phases of pollination, there is also significant evidence that supports a role for NO as an intercellular signal in the final critical events. In *Torenia fournieri*, one of the best-characterized physiological models, semi-*in vivo* experiments showed evidence supporting the production of diffusible signals from the synergid cells of the embryo sac, effective over a few hundred micrometers at most [72]. In a different experimental system using the *maa* mutant of *Arabidopsis*, the ovule entrance (micropyle) can direct pollen tube growth over distances of 50–90 μm [73].

While NO could directly serve the role in directing final pollen guidance, there are also other interesting players that may serve to activate or modify the intracellular NO system present in the pollen tube. GABA (γ-amino butyric acid), a molecule that evolved as a neurotransmitter in animals, has also been proposed to be a part of this navigation system, hypothetically through the formation of a continuous gradient toward the ovule that is sensed by the growing pollen tube [68]. Alandete-Saez *et al.* [74] describe a new plasma membrane protein GEX3 that is expressed in both the male gametophyte and in the egg cell. Their findings show that both overexpression and downregulation of GEX3 block fertilization by altering pollen tube guidance. The importance of the egg apparatus in mycropylar guidance was shown in maize by Márton *et al.* [75], who identified a small secreted protein, ZmEA1, suggested to be responsible for close-range pollen tube guidance. Pollen tube navigation is not restricted to pistil and micropylar guidance. As the growing pollen tube targets an ovule, the terminal stages of communication occur inside the female gametophyte moments before fertilization. Also worth considering is a recently identified mutant in *Arabidopsis* called lorelei. The LRE protein is expressed in synergid cells and affects pollen tube navigation and sperm release inside the embryo sac, while continuing to attract additional pollen tubes [76]. These mechanisms are little understood and go beyond the scope of this chapter (for review see Ref. [58]).

In the Atnos1/AtNOA1 mutant, NO production was first reported to be impaired [65], but this finding was recently reversed in studies that showed that the gene mutant was instead deficient in NO-mediated cGMP conversion [66]. Nevertheless, the Atnoa1/AtNOA1 mutant does have deficiencies in seed set, suggesting that NO/cGMP signaling does participate in *in vivo* pollen tube guidance. Recent experiments have studied *Atnos1* self-pollination and cross-pollinations between *Atnos1/AtNOA1* mutant and wild-type plants [71]. In these crosses, abnormal pollen tube morphology was detected when *Atnos1/AtNOA1* or wild-type pollen tubes grew in *Atnos1/AtNOA1* pistil, but not the reverse. This result indicates that the mutation affects the female tissues and it may explain the reduced seed set observed by Guo *et al.* [65] and confirmed by Moreau *et al.* [66]. The finding that both wild-type and mutant pollens fail to fertilize the *Atnos1/AtNOA1* mutant ovules indicates a possible NO/cGMP ovule signal function that orchestrates the final reorientation of pollen tube penetration into the micropyle (Figure 3.4). Further support for this hypothesis comes from experiments that demonstrated successful phenocopying of the lower seed set of *Atnos1/AtNOA1* when treating the wild-type plants with PTIO, a permeable NO scavenger. Overall, these results indicate that NO/cGMP signaling is a necessary component for successful sexual plant reproduction.

Other recent work supports the role of NO/cGMP signaling in pollen/ovule signaling. The putative contribution of NO in pollen tube micropyle targeting was explored using *kanadi* mutants, which possess defects in floral development and patterning such as production of ectopic ovules and formation of projections of carpel and style on stigmatic tissue from the base of the pistil [77]. Despite these anatomical

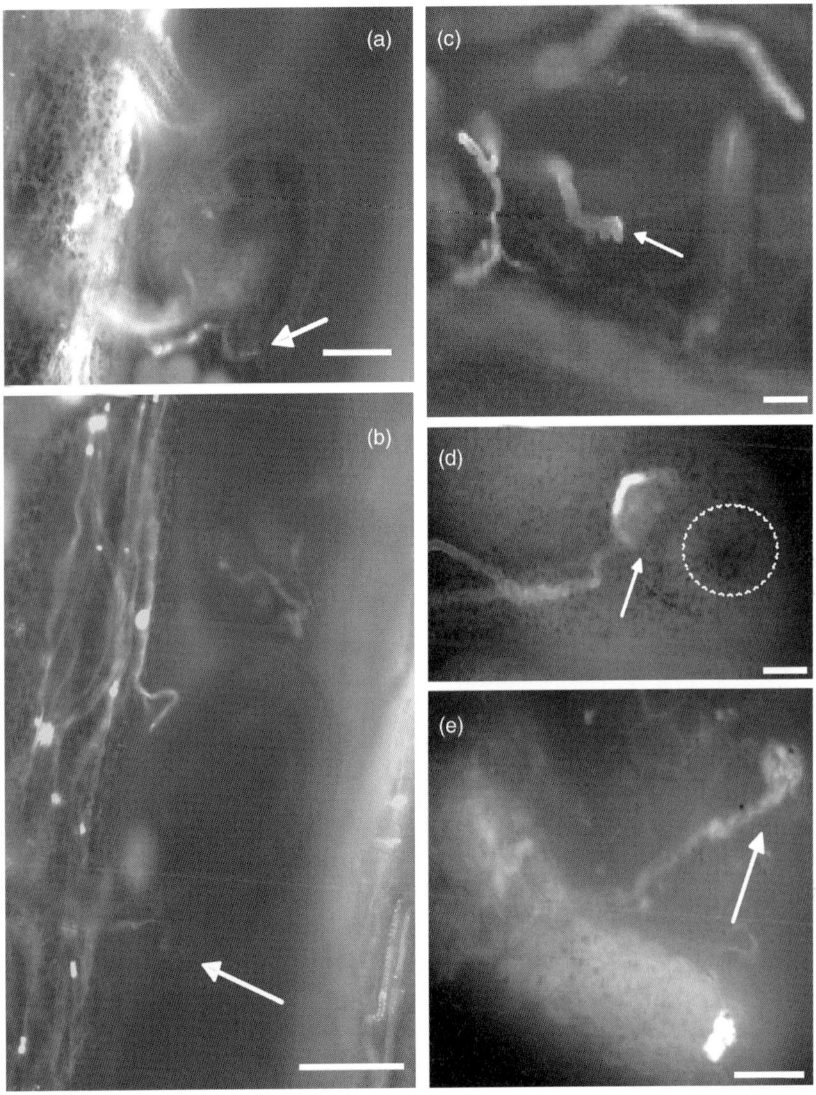

Figure 3.4 Detection of abnormal pollen tube guidance in self- and cross-pollinations between *Atnos1* and Wt plants. Fluorescence microscopy images of Wt and *Atnos1* pistils in self- and cross-pollinations after 6 h. Staining of callose walls with aniline blue. (a) Wt self-pollinated pistil, arrow pointing to Wt pollen tube targeting to the micropyle. Scale bar: 40 μm. (b) Cross-pollination between ♀ Wt × ♂ *Atnos1*, arrow pointing to *Atnos1* pollen tubes at the micropyle entrance. Scale bar: 80 μm. (c) and (d) *Atnos1* self-pollinated pistil. (c) Arrow pointing to abnormal pollen tube tip with comb-like shape; (d) arrow pointing to pollen tube that shows a swollen tip near the micropyle (mp) entrance (white dashed line); (e) cross-pollination between ♀ *Atnos1* × ♂ Wt arrow pointing to the swollen tip of a pollen tube growing on top of an ovule wall. Scale bar: 24 μm [71].

deficiencies, *kanadi* flowers are fertile, and the DAF2-DA probe detected a signal in the micropyle area of *kanadi* flowers. The signal was localized in a restricted number of cells that border the micropyle opening (Figure 3.5). This observation is compatible with targeting of growing pollen tubes to the micropyle being confined by NO negative tropism and reorientation to the micropyle locus through the area deprived of the putative NO signal [71].

Other experiments have used *in vitro* lily ovule culture where the micropyle is arranged such that a row of pollen grains is in the near vicinity. Targeting events can then be observed in the presence and absence of CPTIO, a NO scavenger. These experiments have shown that targeting events decrease in the presence of CPTIO and that pollen tube population was divided into two groups. The first group, located in the top half of the preparation close to the ovules, exhibited tip swelling and cessation of targeting (consistent with high NO exposure or phosphodiesterase inhibition) while the second group in the inferior half of the preparation in the absence of ovules retained polarized growth (Figure 3.5) [71].

These findings support the need for NO in targeting; the distinct behavior of pollen tubes can only be explained by assuming that the ovules produce some sort of diffusible signal that interacts with or needs a basal level of NO to properly function [49]. This is in accordance with the results seen in *Ceratopteris* spore polarity and the hypothesis of combinatorial stimuli as the basis of the specificity with ubiquitous signaling molecules NO and cGMP.

3.7 Ca^{2+}/NO/cGMP Connection?

The question regarding calcium in this system is not yet clear. Does it follow the mammalian eNOS model or is the relationship with calcium different? The loss of

Figure 3.5 (a and b) Identification of putative NO production spots in restricted areas of the micropyle of *Arabidopsis* floral mutants, in exposed ovules labeled with DAF-2DA at the micropyle entrance area. The DAF-2DA signal probes for NO presence. Labeling is observed in both images, being restricted to a cell layer in the immediate vicinity of the micropyle entrance area. Dashed white line shows contour of ovules. (c) *Lilium* pollen tube targeting, showing a curve along the base of the micropyle region at the ovules. (d) Schematic diagram of the semi-*in vivo* preparation to illustrate that the line of pollen divides the preparation into two halves (the frontier between halves is represented by a green line). The top half presents isolated ovules (represented in green) that are aligned with the micropyle facing the row of pollen. Growing pollen tubes are represented by black lines and dots. Black dots represent the balloon pollen tube tips. (e) Cytosolic free calcium ($[Ca^{2+}]_c$) dynamics during NO challenging. *Lilium* pollen tubes were microinjected with Oregon Green BAPTA-Dextran (10 kDa) and imaged in a confocal microscope to visualize the typical tip-focused gradient (e). Tubes growing straight with rates >8 μm min^{-1} (f) were challenged with a SNP-filled pipette source as described in Ref. [48] until the typical reorientation response takes place (g). The plot in (h) shows the alteration in the normalized fluorescence level of the first 15 μm from the apex of a pollen tube during a typical experiment (dots, experimental values; line, moving average, $n = 2$). As the tube enters the diffusion gradient from the pipette, the levels of $[Ca^{2+}]_c$ start to rise (1.5 min on) until reaching

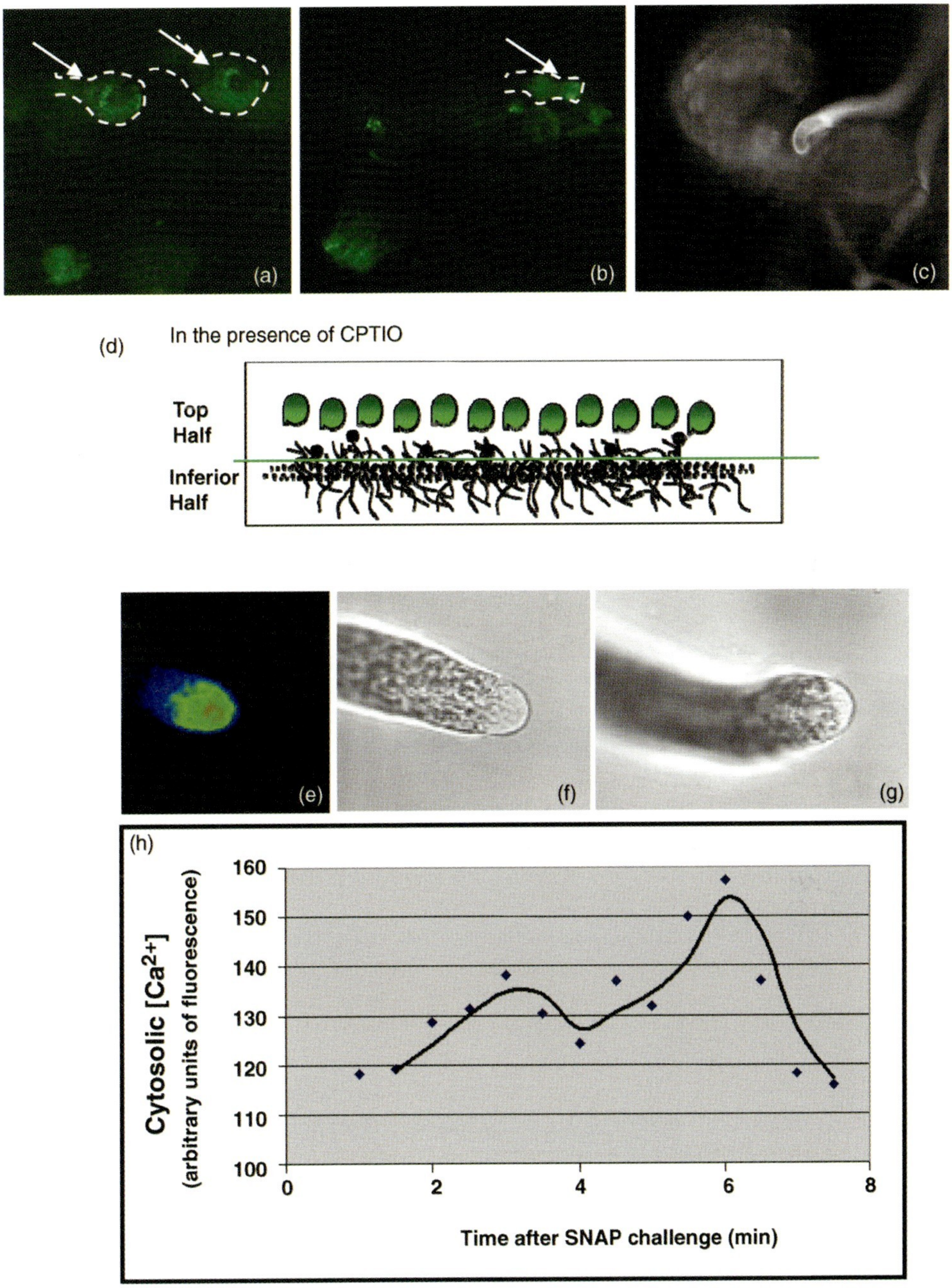

a first peak. This peak coincides with a slowdown, sometimes a full halt, of growth. The levels of $[Ca^{2+}]_c$ then decrease somewhat and start to rise again to a peak that corresponds to the turning point, and the growth restarts (4–6 min). After that, the pollen tubes resume growth at normal rates in the new growth axis (>7 min), and the levels of $[Ca^{2+}]_c$ return to basal levels (here normalized to ~100–110 arbitrary units of fluorescence on an eight-bit scale) [71].

polarized growth by pollen tubes and fern spores is well documented in the literature and well correlated with problems in Ca^{2+} homeostasis in the tip [21, 24, 33, 78]. Therefore, besides the NO signal, in concert with an ovule-derived signal (ODS) factor, we also equate the necessity of a Ca^{2+} gradient at the tip of the pollen tube as an element required for the signaling pathway that drives pollen tube growth and targeting. If this is the case, will cytosolic Ca^{2+} be directly affected by NO? Pollen tubes loaded with Oregon Green BAPTA-Dextran allowed Ca^{2+} imaging during exogenous NO application, suggesting that cytoplasmic NO and Ca^{2+} are interrelated although the exact relationship is still not understood (Figure 3.5).

There are two potential models for how NO and Ca^{2+} interact as part of this signaling cascade. The first model is based on an adaptation of the mammalian eNOS model (Figure 3.1) where calcium-activated calmodulin turns on NO synthase. NO then activates cGMP production that then can modulate ion channel activity. In the model adapted for pollen (Figure 3.6a), the cGMP-gated channel could be a K^+ channel, which alters osmotic potential. Calcium itself responds to the osmotic status of the cell via stretch-activated calcium channels that have been described by Dutta and Robinson [79]. In fact, most of the components for this mechanism have been found in the pollen transcriptome [48]. This model is interesting because it suggests an oscillatory regulatory mechanism where high calcium is opposed by high NO/cGMP yet high calcium activates high NO/cGMP. This is a possible mechanistic model that explains oscillatory growth observed in pollen growth. Hypothetically, this model is comprehensive in terms of regulating the precarious balance between growth and cell rupture that is a key consideration in a fast tip-growing pollen tube.

The alternative hypothesis is that NO precedes Ca^{2+} (Figure 3.6b). While the eNOS model is predominant in mammalian systems, there are also examples of mechanisms where NO regulates the activity of Ca^{2+} channels, namely, the activity of the NMDA receptor in synapses [80]. We must also consider how an ODS modulates growth via a NO-mediated pathway that leads to lily pollen tube targeting. The obvious and simple explanation is that NO itself is the OSD, just as it was discovered to be the long-questioned endothelial relaxing factor in mammals. In this model, NO produced by the ovule simply overrides the endogenous pollen NO signaling system, which is supported by evidence showing NO production in the micropyle.

Still in another version of the model, NO is not ODS but is activated via a receptor. In the semi-*in vivo* preparation, the pollen tubes grow in the presence of ovules and are therefore submitted to a theoretically predicted ovule-derived signal [81]. The ODS receptor could be intracellular, like the NO-producing enzymes (Figure 3.6b), or this could be localized to the pollen membrane (Figure 3.5b). We anticipate that the signaling cascade elicited by the ovule-derived factor could involve NO that could in turn activate Ca^{2+} channels (Figure 3.6b) or inhibit cell expansion via cGMP-gated K^+ channels. In the case of the NO–Ca^{2+} sequence model (Figure 3.6b), a possible physiological sequence for pollen tube targeting would then be (1) ovule-derived signal release, (2) increase in NO at the tip, (3) activation of a putative Ca^{2+} channel at the pollen tube tip, (4) reorientation of the pollen tube mediated by the Ca^{2+} gradient, and (5) redirection of the growth axis.

While these models might not be exclusive, there is still a significant debate on the mechanism of Ca^{2+} entry in plant cells. Frietsch *et al.* [82] recently found a

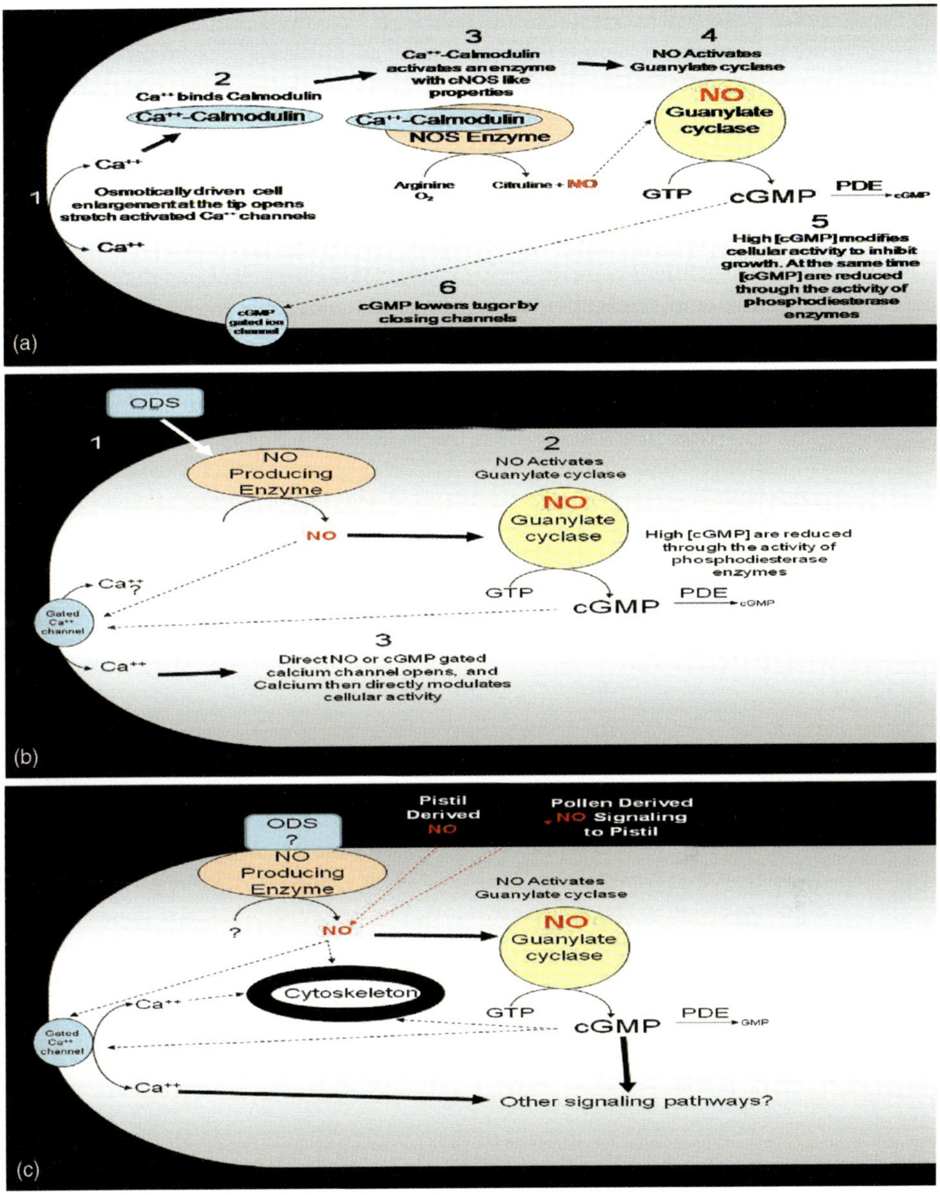

Figure 3.6 Review of some of the models and mechanisms for Ca^{2+} and NO/cGMP signaling in the control of pollen tube growth. Based on an adaptation of the mammalian eNOS system (5a), calcium precedes and activates NO/cGMP signaling in the cell, and based on hypothetical ion channel activities could mechanistically act as an oscillator. Another model (5b) is based on an ovule-derived signal activating NO production, which in turn regulates calcium via direct nitroslyation of a ligand-gated calcium channel or by a cGMP-gated mechanism. The panel in 5c summarizes other significant questions in the pollen tube system, including a role for micropyle-generated NO as a direct signal from the ovule directing pollen tube orientation, and the possibility that pollen-derived NO may act as a signal to the pistil. Other questions include the putative receptor mechanism for the ODS, and a possible direct mechanism for NO regulation of the cytoskeleton via a Cl^- dependent process. And still more complex interactions are possibilities.

remarkable pollen phenotype in pollen for the mutation of a putative cyclic nucleotide-gated channel that was shown to affect Ca^{2+} accumulation when overexpressed in bacteria. Although confirmation of its role as a bona fide Ca^{2+} channel in pollen tubes is still required, that NO may regulate such channels through a cGMP pathway, as suggested in previous experiments [48], remains a tantalizing possibility to link our physiological data and establish a more defined and testable role for NO during pollen–pistil interaction.

Other important questions and considerations are summarized in Figure 3.6c. The question regarding a possible mechanism for controlling pollen growth via ODS and/or NO has been discussed, but what about pollen signaling to the pistil via NO? There are several scenarios where NO and Ca^{2+} may be opposing and regulating one another, but there may be other downstream signaling events that further contribute to this "dance." Another possibility and area of interest involves the control of the cytoskeleton by calcium, NO, and cGMP. A recent paper studying mammalian neutrophylls suggests a direct mechanism for NO regulation of the cytoskeleton via actin *S*-nitrosylation [83]. This mechanism involves a chloride halogenated ROS intermediate, and Cl^- has itself been implicated in pollen tube growth [84].

3.8 Closing Perspectives

Taken together, the pollen and fern spore evidence strongly suggests that both cell polarity and cell tip growth can be regulated by a NO/cGMP signaling pathway in cells where calcium signaling, via a transmembrane calcium current, is a significant signaling mechanism. Both the direction and rate of pollen tube and rhizoid growth are controlled by NO, and in agreement with the animal model (Figure 3.1), by NO-dependent cGMP production. Conservation of a pathway involving a role for NO/cGMP signaling in pterophyte spores, angiosperm pollen, and even in mammalian cardiovascular endothelium is not surprising given the fact that enzymatic NO production has been discovered in bacteria, fungi, protozoans, plants, and animals [85] suggesting that NO may be one of the earliest and most widespread signaling molecules in biological systems. While all of the evidence strongly points to a plant NOS gene that is similar in function and regulation to the animal eNOS, the discovery of this key component has been elusive.

The findings that we present in this section call for a continuing endeavor in understanding the role of NO in pollen–pistil communication. The peculiarity of NO chemistry and the absence of demonstrable genetic mechanisms for its production and transduction raise a justifiable caution to some of the conclusions made [11, 86]. For instance, while *in vivo* NO detection by DAF-2DA has proven to be reliable in many other model systems, NO detection via this technique is not free of artifacts, as DAf-2DA fluorescence is also known to be possible with other more oxidized forms of NO [77]. Nevertheless, it is important to note that the use of this probe in plant biology has revealed robust results, including the detection of asymmetric intracellular NO during gravitropic bending of soybean roots [87].

Addressing the role of NO in pollen–pistil communication in any flowering plant is far from trivial. The necessity of identifying a NO signal in a spatial and temporal context after pollination poses serious technical challenges. The necessity of preventing the production of any lesion-induced NO production in pistil structures needs creative noninvasive approaches that have not been identified and employed to date. Nevertheless, the direct and indirect evidence reviewed here are compatible with the involvement of NO in pollen tube guidance in *Arabidopsis* and lily and ultimately suggest the participation of NO in ovule targeting – the very final step prior to fertilization.

Although our understanding of NO/cGMP activities as a regulator of prefertilization events in flowering plants is just emerging, it is well established as a major player in mammalian reproduction. Miraglia *et al.* [88] point out that NO may act as a chemoattractant in human spermatozoa through a cGMP-dependent pathway (reviewed in Ref. [47]). Moreover, NO is also produced and released by human spermatozoa, suggesting the possibility of an intricate communication to occur among spermatozoa as well as spermatozoa and oocytes. In animals, the stimulation of sperm motility and orientation depends on additional ranges of signals such as peptides, intracellular changes in pH, in the concentrations of cGMP and cAMP, as well as Na^{2+}, Ca^{2+}, and membrane potential, and in the phosphorylation pattern of several proteins (reviewed in Ref. [47]). Do these ions and molecules suggest a common evolutionarily conserved mechanism in both animals and flowering plants?

Most of the aforementioned ions and molecules are already known to regulate pollen tube growth and direct positive chemotropism (cAMP, Ca^{2+}). Calcium has been shown to participate in egg cell activation in maize [89, 90]. It is likely that the relevant mechanisms of cell–cell communication, cell fusion, and prevention of polyspermy evolved prior to the divergence of the separate plant and animal lineage more than 1.5 billion years ago (reviewed in Ref. [46]). Given the biological relevance of fertilization, it is plausible that evolution has created functional redundancy or cofunctionality for different molecules. In fact, theoretical arguments have been raised that a single chemical gradient could hardly be responsible for guidance in most species that led Lush *et al.* [81] to propose mechanical/structural stringencies as cooperative mechanisms in the guidance of pollen tubes. Classical experiments show that the directionality of growth along the pistil/ovary can, in principle, occur in more than one direction, restricting the guidance cue necessity to just a few crucial steps along the pollen tube path and overruling the positive single-molecule chemotropism as the sole mechanism of guidance (reviewed in Refs [45, 58, 91, 92]).

The current knowledge of NO during the evolution of life on Earth contributes to the understanding of NO in biology. It is estimated that life evolved between approximately 3.3–2.4 billion years ago. As photoautotroph cells produced oxygen (O_2), atmospheric levels of O_2 increased as did levels of ozone (O_3), a potent oxidant of biological membranes and organic molecules in general. Detrimental effects of O_3 might have been counteracted by NO. It is conceivable that a mutant might have arisen with the ability to produce NO and to deliver it to the environment (for review see Ref. [93]). The biological consequences of such an evolutionary step must have been far-reaching. As NO does not require a carrier to cross membranes and reach intracellular targets, it is possible that a cellular signaling system between cells could have evolved before the existence of cellular receptors. Regarding the biological origin of NO, it is possible that the pathway for its production derived from mechanisms of denitrification or nitrification. The ubiquity of NO functions from prokaryotic to eukaryotic life organization suggests that NO might have been one of the first biological signaling molecules of life (for review see Ref. [93]). This is only a theory, however, that emphasizes how NO basic function might have been cell–cell communication.

Apart from evolutionary considerations of pollen guidance cues and the primordial function of NO, there is also an all-encompassing question of how generic NO can ultimately impart specificity in plant signaling. Besson-Bard *et al.* [94] addressed this question by highlighting that NO production can be compartmentalized. NO signaling can operate through second messengers with subsequent information propagation controlled in time and space. Therefore, the spatial and temporal arrangements acquired in this manner may determine the specificity of the biological outcome. Moreover, future identification of proteins that can interact with NO-generating enzymes as well as NO target proteins will contribute to deliver NO signal information in a specific fashion.

References

1 Ignarro, J. (2000) Nitric oxide, *Biology and Pathobiology*, Academic Press, San Diego, CA.

2 Crawford, N.M. and Guo, F.Q. (2005) *Trends Plant Sci.*, **10**, 195–200.

3 Delledone, M. (2005) *Curr. Opin. Plant Biol.*, **8**, 390–396.

4 Lamotte, O., Courtois, C., Barnavon, L., Pugin, A., and Wendehenne, D. (2005) *Planta*, **221**, 1–4.

5 Corpas, F.J., Barroso, J.B., Carreras, A., Quiros, M., Leon, A.M., Romero-Puertas, M.C. *et al.* (2004) *Plant Phys.*, **136**, 2722–2733.

6 Lamotte, O., Gould, K., Lecourieux, D., Sequeira-Legrand, A., Lebrun-Garcia, A., Durner, J. *et al.* (2004) *Plant Physiology*, **135**, 516–529.

7 Vandelle, E., Poinssot, B., Wendehenne, D., Bentejac, M., and Pugin, A. (2006) *Mol. Plant–Microbe Interact.*, **19**, 429–440.

8 Sokolovski, S., Hills, A., Gay, R., Garcia-Mata, C., Lamattina, L., and Blatt, M.R. (2004) *Plant J.*, **43**, 520–529.

9 Stamler, J.S. (1994) *Cell*, **78**, 931–936.

10 Lamattina, L., Garcia-Mata, C., Graziano, M., and Pagnussat, G. (2003) *Annu. Rev. Plant Biol.*, **54**, 109–136.

11 Besson-Bard, A., Griveau, S., Bedioui, F., and Wendehenne, D. (2008) *J. Exp. Biol.* doi: 10.1093/jxb/ern189

12 Lamattina, L., Garcia-Mata, C., and Pagnussat, G. (2003) *Ann. Rev. Plant Biol.*, **54**, 109–136.

13 Leshem, Y.Y., Wills, R.B.H., and Veng-Va, K. (1998) *Plant Physiol. Biochem.*, **36**, 825–834.

14 Beligni, M.V. and Lamattina, L. (2000) *Planta*, **210**, 215–221.

15 Mata, C.G. and Lamattina, L. (2001) *Plant Physiol.*, **126**, 1196–1204.

16 Hérouart, D., Baudouin, E., Frendo, P., Harrison, J., Santos, R., Jamet, A., van de Sype, G., Touati, D., and Puppo, A. (2002) *Plant Physiol. Biochem.*, **40**, 619–624.

17 Delledonne, M., Xia, Y., Dixon, R., and Lamb, C. (1998) *Nature*, **394**, 585–588.

18 Durner, J., Wendehenne, D., and Klessig, F. (1998) *Proc. Natl. Acad. Sci. USA*, **95**, 10328–10333.

19 Huang, X., Kiefer, E., von Rad, U., Ernst, D., Foissner, I., and Durner, J. (2002) *Plant Physiol. Biochem.*, **40**, 625–631.

20 Edwards, E.S. and Roux, S.J. (1998) *Planta*, **205**, 553–560.

21 Chatterjee, A., Porterfield, D.M., Smith, P.S., and Roux, S.J. (2000) *Planta*, **210**, 607–610.

22 Stout, S.C., Porterfield, D.M., and Roux, S.J. (2003) *Grav. Space Biol. Bull.*, **17**, 18.

23 ul Haque, A., Rokkam, M., De Carlo, A.R., Wereley, S.T., Wells, H.W., McLamb, W.T., Roux, S.J., Irazoqui, P.P., and Porterfield, D.M. (2007) *Sensor Actuat. B: Chem.*, **123**, 391–399.

24 Holdaway-Clarke, T.L. and Hepler, P.K. (2003) *New Phytol.*, **159**, 539–563.

25 Weisenseel, M.H., Nuccitelli, R., and Jaffe, L.F. (1975) *J. Cell Biol.*, **66**, 556–567.

26 Franklin-Tong, V.E., Holdaway-Clarke, T.L., Straatman, K.R., Kunkel, J.G., and Hepler, P.K. (2002) *Plant J.*, **29**, 333–345.

27 Wang, Y.F., Fan, L.M., Zhang, W.Z., Zhang, W., and Wu, W.H. (2004) *Plant Physiol.*, **136**, 3892–3904.

28 Schiøtt, M., Romanowsky, S.M., Bækgaard, L., Jakobsen, M.K., Palmgren, M.G., and Harper, J.F. (2004) *Proc. Natl. Acad. Sci. USA*, **101**, 9502–9507.

29 Golovkin, M. and Reddy, A.S.N. (2003) *Proc. Natl. Acad. Sci. USA*, **100**, 10558–10563.

30 Malhó, R., Read, N.D., Pais, M.S., and Trewavas, A.J. (1994) *Plant J.*, **5**, 331–341.

31 Malhó, R., Read, N.D., Trewavas, A.J., and Pais, M.S. (1995) *Plant Cell*, **7**, 1173–1184.

32 Malhó, R. and Trewavas, A.J. (1996) *Plant Cell*, **8**, 1935–1949.

33 Pierson, E.S., Miller, D.D., Callaham, D.A., van Aken, J., Hackett, G., and Hepler, P.K. (1996) *Dev. Biol.*, **174**, 160–173.

34 Franklin-Tong, V.E., Hackett, G., and Hepler, P.K. (1997) *Plant J.*, **12**, 1375–1386.

35 Holdaway-Clarke, T.L., Feijó, J.A., Hackett, G.R., Kunkel, J.G., and Hepler, P.K. (1997) *Plant Cell*, **9**, 1999–2010.

36 Messerli, M. and Kinneth, R.R. (1997) *J. Cell Sci.*, **110**, 1269–1278.

37 Camacho, L., Parton, R., Trewavas, A.J., and Malhó, R. (2000) *Protoplasma*, **212**, 162–173.

38 Snowman, B.N., Geitmann, A., Clarke, S.R., Staiger, C.J., Franklin, F.C.H., Emons, A.M.C. *et al.* (2000) *Ann. Bot. (Lond.)*, **85**, 49–57.

39 Camacho, L. and Malhó, R. (2003) *J. Exp. Bot.*, **54**, 83–92.

40 Lazzaro, M.D., Cardenas, L., Bhatt, A.P., Justus, C.D., Phillips, M.S., Holdaway-Clarke, T.L. *et al.* (2005) *J. Exp. Bot.*, **56**, 2619–2628.

41 Franklin-Tong, V.E. (1999) *Curr. Opin. Plant Biol.*, **2**, 490–495.

42 Franklin-Tong, V.E. (1999) *Plant Cell*, **11**, 727–738.

43 Feijó, J.A., Costa, S., Prado, A.M., Becker, J.D., and Certal, A.C. (2004) *Curr. Opin. Plant Biol.*, **7**, 589–598.

44 Hepler, P.K. (2005) *Plant Cell*, **17**, 2142–2155.

45 Boavida, L.C., Vieira, A.M., Becker, J.D., and Feijó, J.A. (2005) *Int. J. Dev. Biol.*, **49**, 615–632.

46 Márton, M.L. and Dresselhaus, T. (2008) *Sex. Plant Reprod.*, **21**, 37–52.

47 Higashiyama, T. and Hamamura, Y. (2008) *Sex. Plant Reprod.*, **21**, 17–26.

48 Prado, A.M., Porterfield, D.M., and Feijó, J.A. (2004) *Development*, **131**, 2707–2714.

49 McInnis, S.M., Desikan, R., Hancock, J.T., and Hiscock, S.J. (2006) *New Phytol.*, **172**, 221–228.

50 Michard, E., Alves, F., and Feijó, J.A. (2009) Int. J. Dev. Biol., doi:10.1387/ijdb.072296em.

51 Salmi, M.L., Morris, K.E., Roux, S.J., and Porterfield, D.M. (2007) *Plant Physiol.*, **144**, 94–104.

52 Becker, J.D., Boavida, L.C., Carneiro, J., Haury, M., and Feijó, J.A. (2003) *Plant Physiol.*, **133**, 713–725.

53 Pina, C., Pinto, F., Feijó, J.A., and Becker, J.D. (2005) *Plant Physiol.*, **138**, 744–756.

54 Pruitt, R.E. (1999) *Curr. Opin. Cell Biol.*, **2**, 419–422.

55 Palanivelu, P. and Preuss, D. (2000) *Trends Cell Biol.*, **10**, 517–524.

56 Cheung, A.Y. and Wu, H.M. (2001) *Science*, **293**, 1441–1442.

57 Johnson, S.A. and Preuss, D. (2002) *Dev. Cell*, **2**, 273–281.

58 Lord, E.M. and Russell, S.D. (2002) *Annu. Rev. Cell Dev. Biol.*, **18**, 81–105.

59 Wolters-Arts, M., Lush, W.M., and Mariani, C. (1998) *Nature*, **392**, 818–821.

60 Mascarenhas, J.P. and Machlis, L. (1962) *Nature*, **196**, 292–293.

61 Mascarenhas, J.P. and Machlis, L. (1964) *Plant Physiol.*, **39**, 70–77.

62 Wu, H.M., Wang, H., and Cheung, A.Y. (1995) *Cell*, **82**, 395–403.

63 Mollet, J.C., Park, S.Y., Nothnagel, E.A., and Lord, E.M. (2000) *Plant Cell*, **12**, 1737–1749.

64 Boisson-Dernier, A., Frietsch, S., Kim, T.-H., Dizon, M.B., and Schroeder, J.I. (2008) *Curr. Biol.*, **18**, 63–68.

65 Guo, F.Q., Okamoto, M., and Crawford, N.M. (2003) *Science*, **302**, 100–103.

66 Moreau, M., Lee, G.I., Wang, Y., Crane, B.R., and Klessig, D.F. (2008) *J. Biol. Chem.*, **283**, 32957–32967.

67 Wheeler, M.J., Franklin-Tong, V.E., and Franklin, F.C.H. (2001) *New Phytol.*, **151**, 565–584.

68 Palanivelu, R., Bras, L., Edlund, A.F., and Preuss, D. (2003) *Cell*, **114**, 47–59.

69 Bright, J., Desikan, R., Hancock, J.T., Weir, I.S., and Neill, S.J. (2006) *Plant J.*, **45**, 113–122.

70 Boavida *et al.* (2008), submitted for publication.

71 Prado, A.M., Colaço, R., Moreno, N., Silva, A.C., and Feijó, J.A. (2008) *Mol. Plant.* doi: 10.1093/mp/ssn034

72 Higashyama, T., Kuroiwa, H., Kawano, S., and Koroiwa, T. (2003) *Curr. Opin. Plant Biol.*, **6**, 36–41.

73 Shimizu, K.K. and Okada, K. (2000) *Development*, **127**, 4511–4518.

74 Alandete-Saez, M., Ron, M., and McCormick, S. (2008) *Mol. Plant.* doi: 10.1093/mp/ssn014

75 Márton, M.L., Cordts, S., and Broadhvest, J. Dresselhaus. (2005) *J. Sci.*, **307**, 573–576.

76 Capron, A., Gourgues, M., Neiva, L.S., Faure, J.-E., Pagnussat, G., Krishnan, A., Alvarez-Mejia, C., Vielle-Calzada, J.-P., Lee, Y.-R., Liu, B., and Sundaresan, V. (2008) *Plant Cell*, **20**, 3038–3049.

77 Kerstetter, R.A., Bollman, K., Taylor, R.A., Bomblies, K., and Poething, R.S. (2001) *Nature*, **411**, 706–709.

78 Michard, E., Alves, R.F., and Feijó, J.A. (2008) *Int. J. Dev. Biol.*, in press.

79 Dutta, R. and Robinson, K.R. (2004) *Plant Physiol.*, **135**, 1398–1406.

80 Boehning, D. and Snyder, S.H. (2003) *Annu. Rev. Neurosci.*, **26**, 105–131.

81 Lush, W.M., Grieser, F., and Wolters-Arts, M. (1998) *Plant Physiol.*, **118**, 733–741.

82 Frietsch, S., Wang, Y.F., Sladek, C., Poulsen, L.R., Romanowsky, S.M., Schroeder, J.I., and Harper, J.F. (2007) *Proc. Natl. Acad. Sci. USA*, **104**, 14531–14536.

83 Thom, S.R., Bhopale, V.H., Mancini, D.J., and Milovanova, T.N. (2008) *J. Biol. Chem.*, **283**, 10822–10834.

84 Zonia, L., Cordeirob, S., Tupa, J., and Feijó, J.A. (2002) *Plant Cell*, **14**, 2233–2249.

85 Torreilles, J. (2001) *Front. Biosci.*, **6**, D1161–D1172.

86 Planchet, E. and Kaiser, W.M. (2006) *Plant Signal. Behav.*, **1**, 46–51.

87 Hu, X., Neill, S.J., Tang, Z., and Cai, W. (2005) *Plant Physiol.*, **137**, 663–670.

88 Miraglia, E., Rullo, M.L., Bosia, A., Massobrio, M., Revelli, A., and Dario, G. (2007) *Fertil. Steril.*, **87**, 1059–1063.

89 Antoine, A.-F., Faure, J.E., Cordeiro, S., Dumas, C., Rougier, M., and Feijo, J.A. (2000) *Proc. Natl. Acad. Sci. USA*, **19**, 10643–10648.

90 Antoine, A.-F., Faure, J.E., Dumas, C., and Feijo, J.A. (2001) *Nat. Cell Biol.*, **3**, 1120–1123.

91 Heslop-Harrison, J. (1987) *Int. Rev. Cytol.*, **107**, 159–187.

92 Mascarenhas, J.P. (1993) *Plant Cell*, **5**, 1303–1314.

93 Feelisch, M. and Martin, J.F. (1995) *Trends Ecol. Evol.*, **10**, 496–499.

94 Besson-Bard, A., Pugin, A., and Wendehenne, D. (2008) *Annu. Rev. Plant Biol.*, **59**, 21–39.

4
Nitric Oxide and Abiotic Stress in Higher Plants

Francisco J. Corpas, José M. Palma, Marina Leterrier, Luis A. del Río, and Juan B. Barroso

Summary

Nitric oxide (NO) is an endogenous component in plant cell metabolism that is involved in multiple physiological processes such as germination, root development, and senescence. NO and related molecules designated as reactive nitrogen species experience a complex metabolism in cells affecting numerous molecules involved in signaling and stress processes. In plants under adverse and stress conditions, the cell homeostasis of nitric oxide and NO-derived molecules can be deregulated. This imbalance may cause multiple damages to different cellular components including proteins, lipids, and nucleic acids and results in nitrosative stress. In this chapter, the current knowledge of NO behavior in plants exposed to diverse environmental stress conditions such as salinity, heavy metals, UV-B radiation, ozone, and mechanical wounding will be reviewed.

4.1
Introduction

Nitric oxide (NO) is a gaseous free radical with important physiological functions in higher plants [1–10]. Even though much significant knowledge has been gained in the past decade regarding NO and its interactions with other molecules, it must be realized that there are still more questions about NO than clear answers. A good example is the identification of the responsible NO-generating enzyme/s in plant cells under normal or stress conditions, still a controversial issue [2, 7, 8, 11–13]. On the other hand, recent data on the interaction of NO or NO-derived molecules with proteins, lipids, and nucleic acids have revealed the presence of an important mechanism of redox-based modifications that can regulate multiple processes under physiological and stress conditions.

Nitric Oxide in Plant Physiology. Edited by S. Hayat, M. Mori, J. Pichtel, and A. Ahmad

ISBN: 978-3-527-32519-1

4.2
Nitric Oxide and Related Molecules

4.2.1
Chemistry of Nitric Oxide in Plant Cells

Nitric oxide is a simple molecule, but its biochemistry has been shown to be quite complex, and there are a variety of challenging characteristics that have not been fully understood. NO is a lipophilic radical that diffuses across cell membranes and through the cytoplasm at a rate of 50 μm per second and whose solubility in aqueous solutions is 1.9 mM at 1 atm pressure. The *in vivo* NO lifetime is relatively short, that is, less than 10 s [14]. Nitric oxide can impart dual effects depending on cellular concentrations. Low concentrations (nanomolar) are required for regulatory processes; however, high concentrations of NO (micromolar) are involved in multiple damages to different cellular components including proteins, lipids, and nucleic acids.

In aqueous solution, NO reacts with oxygen to form nitrogen dioxide (NO_2) (Eq. (4.1)) with a rate constant of $6.3 \times 10^6\ M^{-2}\ s^{-1}$. Even when some controversial intermediates occur, the reaction continues to finally yield nitrite (NO_2^-) (Eqs. (4.2) and (4.3))

$$2^{\bullet}NO + O_2 \rightarrow 2^{\bullet}NO_2 \quad (4.1)$$

$$^{\bullet}NO_2 + {}^{\bullet}NO \rightarrow N_2O_3 \quad (4.2)$$

$$N_2O_3 + H_2O \rightarrow 2NO_2^- + 2H^+ \quad (4.3)$$

NO has also the capacity to react with other radicals such as superoxide radical (O_2^-) yielding peroxynitrite (Eq. (4.4)) with a rate constant of approximately $6.7 \times 10^9\ M^{-1}\ s^{-1}$. This molecule is a potent oxidant and can react with most types of biomolecules *in vitro* [14].

$$^{\bullet}NO + O_2^{\bullet -} \rightarrow ONOO^- \quad (4.4)$$

In the cell environment, NO also reacts with reduced thiols to form *S*-nitrosothiols (RSNO). This reaction is designated as *S*-nitrosylation and can affect proteins, peptides, and amino acids (Figure 4.1). The reaction of NO with glutathione (GSH) forms *S*-nitrosoglutathione (GSNO), which has biological significance due to its abundance within various subcellular compartments in plant cells. These compounds act as NO donors and can therefore mediate the NO effect in biological systems. RSNOs release NO very slowly, but this process can by accelerated by Cu^+ *in vitro* [15].

4.2.2
Reactive Nitrogen Species

The term "reactive nitrogen species" (RNS) have been formulated to designate nitric oxide and NO-derived molecules such as *S*-nitrosothiols, *S*-nitrosoglutathione

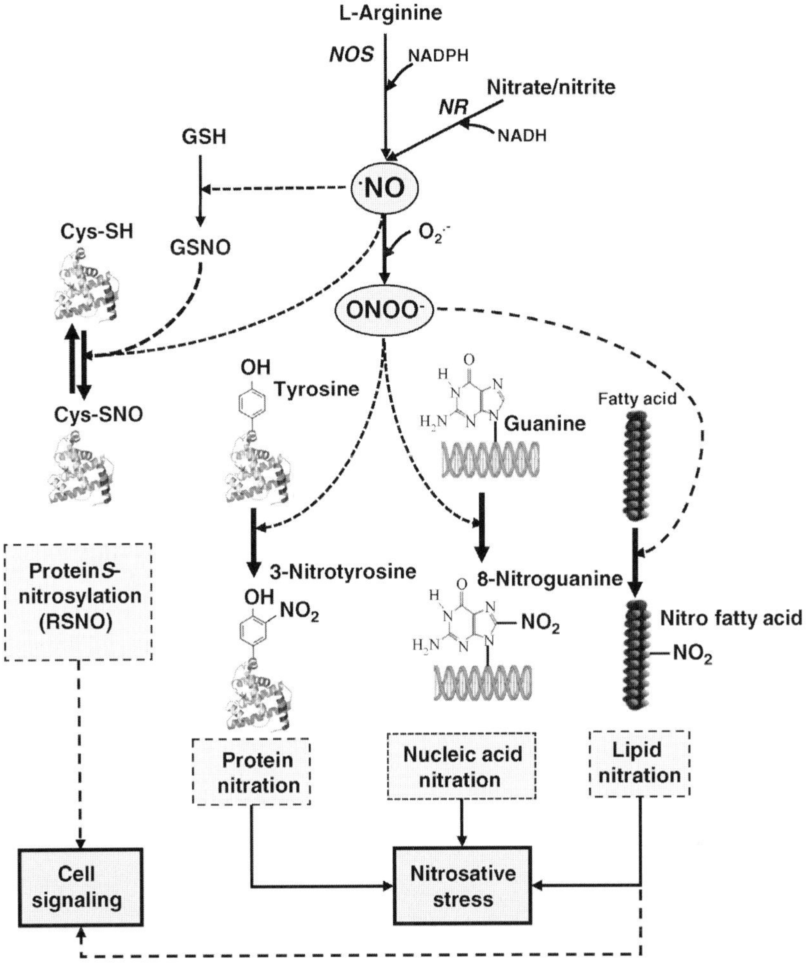

Figure 4.1 NO and other reactive nitrogen species can mediate modifications of biomolecules. Some of these modifications can be involved in plant cell signaling (protein *S*-nitrosylation) or in the process of nitrosative stress. L-Arginine-dependent nitric oxide synthase and nitrate reductase generate $NO^{\bullet}$ that can react with reduced glutathione (GSH) in the presence of O_2 to form *S*-nitrosoglutathione (GSNO), which is the most abundant RSNO. On the other hand, $NO^{\bullet}$ can react with superoxide radicals ($O_2^{\bullet-}$) to generate peroxynitrite ($ONOO^-$), a powerful oxidant molecule that can mediate modifications in biomolecules (proteins, nucleic acids, and lipids). Some of these modifications may be involved in cell signaling (protein *S*-nitrosylation or lipid nitration). However, overproduction of these species under stress conditions can raise the process of nitration in proteins (3-nitrotyrosine), nucleic acids (8-nitroguanine), and fatty acids that could be considered potential markers of nitrosative stress.

(GSNO), and peroxynitrite ($ONOO^-$). A summary of NO and other RNS (radical and nonradical) is presented in Table 4.1. These molecules have relevant roles in multiple physiological and pathological processes in both animal and plant cells [3, 16]. Little is known about RNS in plant cells compared to those in animal systems. One reason

Table 4.1 NO-derived molecules designated as reactive nitrogen species including radical and nonradical molecules.

Radicals	No radicals
Nitric oxide ($^{\bullet}NO$)	Nitrous acid (HNO_2)
Nitrogen dioxide ($^{\bullet}NO_2$)	Dinitrogen trioxide (N_2O_3)
	Dinitrogen tetraoxide (N_2O_4)
	Nitronium (nitryl) ion (NO_2^+)
	Peroxynitrite ($ONOO^-$)
	Peroxynitrous acid (ONOOH)
	Alkyl peroxynitrites (ROONO)
	Nitroxyl anion (NO^-)
	Nitrosonium cation (NO^+)

may be related to the difficulty in detecting these RNS in biological samples because they have very short lifetimes as a consequence of their reactivity with other molecules. However, RNS must be taken into account because they may act as a reservoir and transporter of NO in a variety of cell signaling events.

4.3 Cellular Targets of NO

As mentioned above, the biochemistry of NO allows to establish a complex network in cell metabolism, with many pathways not well defined in plants. Nitric oxide or NO-derived reactive nitrogen species can mediate specific modifications in biological macromolecules that affect their functions, so they could be used as indicators of physiological dysfunction under stress conditions and as instruments to regulate cell metabolism.

4.3.1 Nitrosylated Metals

Plants possess a variety of metalloproteins containing transition metals in their structure that bind to or react with NO (Eq. (4.5)). Metal–nitrosyl complexes are formed under neutral physiological conditions and are proposed to act as a link between different redox states of NO [17].

$$^{\bullet}NO + Fe^{2+} \rightarrow Fe^{2+}\text{-}NO \qquad (4.5)$$

The most studied NO complex involves NO binding to the heme iron in a significant number of proteins such as the Fe storage protein ferritin; heme-containing proteins (guanylate cyclase, cytochrome P450); [Fe–S] cluster proteins involved in energy metabolism such as mitochondrial aconitase; and those proteins from complex I and II of the mitochondrial electron transport chain [18, 19].

4.3.2
Protein S-Nitrosylation

S-Nitrosylation involves the binding of a NO group to a cysteine (Cys) residue of a protein; this product can alter the function of a broad spectrum of proteins (Figure 4.1). In plants, using proteomic approaches, some putative protein targets for *S*-nitrosylation have been identified including cytoskeleton, metabolic, redox-related, stress-related, and signaling/regulating proteins [20, 21]. However, for few proteins, there is experimental evidence showing that they are regulated by this posttranslational modification. Thus, in *Arabidopsis*, nonsymbiotic hemoglobin (AHb1) scavenges NO with production of *S*-nitrosohemoglobin and reduces NO emission under hypoxic stress; glyceraldehyde 3-phosphate dehydrogenase suffers a reversible inhibition by NO [20]; and methionine adenosyltransferase is inhibited by *S*-nitrosylation [20].

4.3.3
Protein Tyrosine Nitration

Protein tyrosine nitration consists of an addition of a nitro group ($-NO_2$) to one of the two equivalent *ortho*-carbons of the aromatic ring of tyrosine residues. This process appears to be mediated by peroxynitrite (Figure 4.1). In animal cells, protein tyrosine nitration has been used as a biomarker of pathological states and nitrosative stress because it modifies the conformation and structure of proteins, the catalytic activity of enzymes, and their susceptibility to proteolysis. In plants, although less information is available, recent data indicate that nitrotyrosine could also be used as a marker of nitrosative stress [22]. Thus, in suspension cultures of *Taxus cuspidate*, an increase of 31% in the free 3-nitrotyrosine content during shear stress has been reported [23]. In nitrite reductase antisense tobacco leaves, the induction of several tyrosine-nitrated polypeptides with molecular masses between 10 and 50 kDa has also been described [24]. Moreover, in tobacco BY-2 suspension cells treated with a fungal elicitin, the induction of tyrosine nitration in proteins with molecular masses in the range 20–50 kDa was demonstrated [25]. All these data indicate that an increase in the number of proteins or an intensification of specific proteins resulting from tyrosine nitration could be considered an indicator of nitrosative stress in plants, as has been demonstrated in animal cells [26].

4.3.4
Nitrolipids

Fatty acid nitration is a recently discovered process that generates biologically active nitrolipids that could be involved in signaling or pathological processes [27, 28]. In animal systems, it has been shown that nitrated membranes and lipids from lipoproteins can transduce NO-signaling reactions and mediate regulatory pathways for anti-inflammatory processes [29]. For example, nitroalkene isomer derivatives of linoleic acid (LNO_2) and oleic acid ($OA\text{-}NO_2$) have been detected in human blood.

These derivatives have the capacity to regulate gene expression and PPAR (peroxisome proliferator-activated receptor) activation due to their electrophilic reactivity [22]. However, higher levels of nitrolipids may serve as novel biomarkers in pathophysiological processes.

Research in nitrolipids is still a new area of investigation and, to our knowledge, there is virtually no information available on nitrolipids in plant systems.

4.3.5
Nucleic Acid Nitration

Nucleotides within DNA and RNA can experience nitration by various RNS (e.g., peroxynitrite and nitrogen dioxide) with the formation of 8-nitroguanine (Figure 4.1) [30]. In animal systems, 8-nitroguanine has been found to act as a prooxidant to stimulate superoxide generation by NADPH cytochrome P450 reductase and nitric oxide synthases. Once formed in cells, 8-nitroguanine may impart pathophysiological consequences due to its mutagenic and prooxidant properties, as well as serving as a footprint of biological nitration [31, 32].

4.3.6
NO and Gene Regulation

Nitric oxide can regulate gene expression in yeast, mammals, and plants [33, 34] through different pathways, either directly or indirectly [35, 36]. In mammalian cells, NO is involved in gene silencing. Thus, endogenously produced or exogenously added NO activates a DNA methyltransferase, leading to methylation of cytosine within the fragile X mental retardation gene *FMR1* followed by consequent suppression of the *FMR1* mRNA [37]. NO also acts at the posttranscriptional level in ferritin synthesis. Translation of ferritin mRNA is repressed by the binding of two *trans*-acting factors, IRP1 (iron regulatory protein) and IRP2, to the 5′-untranslated mRNA region [38]. Exposure to NO disassembles the iron–sulfur cluster of IRP1 through metal nitrosylation promoting the binding of the ferritin mRNA [38, 39]. In contrast, IRP2 binding is inhibited by *S*-nitrosylation of the protein and leads to its degradation by the ubiquitin/26S proteasome pathway [40, 41]. It is interesting to note that in plants, ferritin synthesis is also regulated by NO and the ubiquitin-dependent protein degradation pathway, but at the transcriptional level [42]. Nevertheless, most of the works published thus far regarding gene regulation by NO concern posttranslational modifications of transcription factors (TFs) [35, 36]. NO and NO-derived species can interact directly with TFs through *S*-nitrosylation of cysteine residues, tyrosine nitration, and metal nitrosylation.

Numerous studies on the NF-κB (nuclear factor-kappa B) demonstrate the complexity of NO-regulation. For example, NO can activate or inhibit NF-κB-dependent gene transcription depending on the molecular target, that is, *S*-nitrosylation of the p50 subunit, nitration of the IκBα (inhibitor of kappa B alpha), and the NO-derived species (SNP, endogenous NO) [43–46].

Many TFs are regulated by *S*-nitrosylation, for example, OxyR in bacteria [47], AP-1 [48], and c-Myb [49] in mammals, and AtMYB2 in *Arabidopsis* [50]. In the last case, it has been shown by pharmacological approaches that *S*-nitrosylation of Cys53 residue of the AtMYB2 transcription factor inhibits its binding to DNA [50]. This transcription factor is a typical R2R3-MYB from *Arabidopsis thaliana* that regulates a variety of processes including responses to environmental factors and hormones, developmental processes and cell fate, and metabolic pathways.

Information on TFs tyrosine nitration is scarce; however, Cobbs *et al.* [51] demonstrated that the p53 tumor suppressor TF activity is regulated by nitration in malignant glioma cells.

The ability of NO to react with the metal center in metalloproteins is another way of regulating TF activity in bacteria [52], yeast [53], and mammals. In fact, NO destroys [ZnS] clusters leading to inhibition of zinc finger–TFs binding with DNA [54].

Recently, efforts have been geared toward identification of nitric oxide responsive elements (NORE) in yeast [55] and in promoters of the entire *Arabidopsis* genome by bioinformatic methods [56]. In this species, some *cis* elements corresponding to eight TFs families (WRKY, GBOX, OCSE, etc.) have been found to be 15–28% more represented in the promoter region of genes regulated by NO [56]. Unfortunately, although this work provides insight into TF families more likely to be NO-regulated, it does not identify potential targets within these families and does not account for indirect regulation of gene expression as, for example, nitration of proteins modulating TF activity [46].

4.4 Functions of NO in Plant Abiotic Stress

In plant cells, NO and NO-derived molecules are involved in response to many abiotic stresses such as salinity, drought, ozone, heavy metal toxicity, extreme temperature, ultraviolet radiation, mechanical wounding, and so on. Consequently, when a specific abiotic stress alters physiological NO metabolism causing damage to biological molecules, a nitrosative stress is generated [3, 57].

4.4.1 Salinity

Salinity affects plant productivity due of its negative effects on plant growth, ion balance, and water relations. In many plant species, it has been shown that NaCl provokes an oxidative stress [58–61]; however, less is known about NO metabolism.

In olive plantlets, salt stress (200 mM NaCl) caused an increase in the L-arginine-dependent production of NO, total *S*-nitrosothiols (RSNO), and number of proteins that underwent tyrosine nitration in the molecular mass range between 44 and 60 kDa [57]. Moreover, confocal laser scanning microscopy (CLSM) analysis using either specific fluorescent probes for NO and RSNO or antibodies to *S*-nitrosoglutathione and 3-nitrotyrosine also showed a general increase in these reactive nitrogen species, mainly in the vascular tissue [57]. These findings indicate that in olive leaves salinity induces nitrosative stress, and vascular tissues could play an important role in the redistribution of NO-derived molecules during nitrosative stress.

There is also information on the application of NO donors to different plants and their response to salt stress. In maize, addition of exogenous NO increases tolerance to salt stress by elevating the activities of the proton-pump and Na^+/H^+ antiport of the tonoplast [62]. In *Lupinus luteus*, 200 mM NaCl inhibits germination, but preincubation of seeds with the NO donor sodium nitroprussiate (SNP) restored germination [63].

In *A. thaliana*, the effect of NaCl on wild-type and *Atnoa1* mutant (with defect in *in vivo* NOS activity) has been studied [64]. The authors observed that salt stress (300 mM NaCl) inhibited NOS activity and reduced endogenous NO levels in *Arabidopsis* wild-type plants. On the other hand, addition of NO donors (SNP) alleviated salt stress symptoms whereas the NOS inhibitor (L-NNA) and the NO scavenger (PTIO) exacerbated salt stress effects. Nevertheless, these results contrast the previous observations in olive plants [57].

4.4.2 Ultraviolet Radiation

Incoming UV-B radiation (280–320 nm) has increased in recent years as a consequence of damage to the stratospheric ozone layer, and this radiation clearly affects plant growth and usually induces oxidative stress. In *A. thaliana*, a NOS inhibitor and a NO scavenger partially blocked the induction of the chalcone synthase gene by UV-B [65]. In maize seedlings, UV-B radiation strongly induced NOS activity and decreased both leaf biomass and *exo*- and *endo*-β-glucanase activities. In bean seedlings subjected to UV-B radiation, exogenous NO partially alleviated the UV-B effect characterized by a decrease in chlorophyll content and oxidative damage to the thylakoid membrane [66]. Moreover, UV-B induced stomatal closure, which was mediated by NO and H_2O_2, and the generation of NO was caused by a NOS-like activity [67]. However, other authors reported that NO generated in guard cells were produced by nitrate reductase (NR) activity [68].

4.4.3 Ozone

Ozone (O_3) is formed in the troposphere when sunlight induces photochemical reactions involving oxides of nitrogen (NO_x) and volatile hydrocarbons. The O_3 molecule penetrates plant leaves through stomata during normal gas exchange and causes varying symptoms including chlorosis and necrosis. Pharmacological approaches have revealed that O_3 treatment of *Arabidopsis* plants induced NOS activity that preceded accumulation of salicylic acid and subsequent cell death. In addition, NO treatment has been shown to increase levels of ozone-induced ethylene production and increase leaf injury [69]. In tobacco plants, fumigated with O_3, the accumulation of hydrogen peroxide in mitochondria and early accumulation of NO and ethylene in leaf tissues have been described. The accumulated NO was involved in the induction of alternative oxidase (*AOX*) gene expression, which serves as a mechanism to counteract oxidative stress [70].

Isoprene is one of the most abundant volatile organic compounds emitted by plants. Using leaves of *Phragmites* exposed to ozone, it has been shown that endogenous isoprene was able to decrease the concentration of tissue NO [71].

4.4.4
Mechanical Wounding

Mechanical wounding in plants is a consequence of diverse environmental stresses such as wind, sand, rain, or herbivore attack, among others. In general, plants respond via the induction of numerous genes because the open wound could be a potential infection site for pathogens. Therefore, the expression of defense genes at the wound site is a barrier against opportunistic microorganisms [72, 73].

In tomato plants, the application of NO donors such as sodium nitroprusside or *S*-nitroso-*N*-acetyl-penicillamine (SNAP) inhibited the expression of wound-inducible proteinase inhibitors [74]. In *A. thaliana*, mechanical wounding induced a rapid accumulation of NO that could be involved in jasmonic acid-associated defense responses and adjustments [75]. In pea leaves, mechanical wounding also induced an accumulation of NO after 4 h that was accompanied by a general induction of NOS and GSNOR activities, as well as an increase in the content of RSNOs [76]. However, the pattern of proteins that underwent tyrosine nitration was not affected.

In the wound-healing response of potato leaves, it has been shown that application of the NO donor sodium nitroprusside induced the deposition of the cell-wall glucan callose. Moreover, this was accompanied by an increased expression of wound-related genes such as phenylalanine ammonia-lyase (*PAL*) and extensin [77].

4.4.5
Toxic Metals (Cadmium and Aluminum)

Cadmium (Cd^{2+}) is a highly toxic heavy metal that strongly adsorbs to organic matter in soils; such adsorption is dangerous, as Cd uptake through the food chain will increase. Soils that are acidified enhance cadmium uptake by plants. This effect is a potential danger to animals that depends on Cd-rich plants for survival. For instance, cattle may accumulate significant amounts of cadmium in their kidneys, especially when they ingest multiple plant species [78]. As a result, cadmium is a toxic pollutant to humans, animals, and plants as components of the trophic chain [67, 79]. In pea plants exposed to 50 μM $CdCl_2$, growth was significantly inhibited; in addition, a reduction in transpiration and photosynthesis rates and chlorophyll content was detected, along with imbalances in plant nutrient status [80, 81]. Moreover, Cd^{2+} caused disturbances in plant antioxidant defenses and induced oxidative stress in leaves and roots of pea plants [81, 82]. The contents of nitric oxide, glutathione (GSH), and GSNO significantly decreased and were accompanied by a concomitant reduction in *S*-nitrosoglutathione reductase (GSNOR) activity and transcript expression in leaves [83]. In soybean plants exposed to an acute level of $CdCl_2$ (200 µM), similar effects on growth parameters were observed, but the exogenous application of nitric oxide protected against oxidative damage caused by this metal stress, increasing levels of heme oxygenase-1 expression, as it occurs with other genes involved in the antioxidant defense system [84]. In contrast, pretreatment of seedlings with 100 µM SNP protected sunflower leaves against Cd-induced oxidative stress [85]. A similar effect has been described in *Lupinus* roots grown with 50 µM Cd^{2+}, and it was proposed that the protective effect of NO could consist of stimulation of superoxide dismutase activity to counteract overproduction of superoxide radicals, thus avoiding formation of peroxynitrite from NO and O_2^- [63].

Aluminum (Al^{3+}) is another major factor limiting crop growth and yield in acidic soils because it inhibits cell division, cell extension, and transport [86, 87]. *Hibiscus moscheutos* exposed to 100 µM $AlCl_3$ experienced inhibition of root growth. This effect was accompanied by inhibition of NOS activity and reduced NO concentrations [88]. On the other hand, *Cassia tora* plants pretreated for 12 h with 0.4 mM SNP, a NO donor, and subsequently exposed to 10 µM aluminum treatment for 24 h exhibited significantly greater root elongation and a decrease in Al^{3+} accumulation in root apexes compared to plants without SNP treatment [89].

In *A. thaliana* roots, cells of the distal portion of the transition zone emitted large quantities of NO, which was blocked by 90 µM aluminum treatment [90].

4.5 Concluding Remarks

There is a considerable lack of knowledge regarding the function of NO and other RNS in the biochemistry and physiology of plants under stress conditions. For example, it is now known that target molecules including proteins, nucleic acids, and lipids that, depending of their levels, can be either regulatory molecules in signal communications or can cause damage resulting in cell death. Thus, in plant research many challenges await solution, such as the identification of the enzymatic source/s of NO in plants as well the cellular and subcellular localization of this enzymatic source, and whose regulation will determine its physiological significance. Moreover, there are other open questions as follows: How can nitric oxide/RSNO function as a specific and reversible signal molecule? Which mechanisms in biological molecules determine whether they promote positive or detrimental effects? Additional research is necessary to address all these questions.

Acknowledgments

This work was supported by grants from the Ministry of Education and Science (BIO2006-14949-C02-01 and BIO2006-14949-C02-02) and Junta de Andalucía (Agencia de Innovación y Desarrollo de Andalucía, Consejería de Innovación, Ciencia y Empresa, and research groups CVI 192 and CVI 286).

References

1 Corpas, F.J., del Río, L.A., and Barroso, J.B. (2008) *Plant Signal. Behav.*, **3**, 301–303.

2 Corpas, F.J., Barroso, J.B., Carreras, A., Valderrama, R., Palma, J.M., León, A.M., Sandalio, L.M., and del Rio, L.A. (2006) *Planta*, **224**, 246–254.

3 Corpas, F.J., Carreras, A., Valderrama, R., Chaki, M., Palma, J.M., del Río, L.A., and Barroso, J.B. (2007) *Plant Stress*, **1**, 37–41.

4 Corpas, F.J., Barroso, J.B., and del Rio, L.A. (2007) *Trends Plant Sci.*, **12**, 436–438.

5 Lamattina, L., García-Mata, C., Graziano, M., and Pagnussat, G. (2003) *Annu. Rev. Plant Biol.*, **54**, 109–136.

6 Neill, S.J., Desikan, R., and Hancock, J.T. (2003) *New Phytol.*, **159**, 11–35.

7 Neill, S., Bright, J., Desikan, R., Hancock, J., Harrison, J., and Wilson, I. (2008) *J. Exp. Bot.*, **59**, 25–35.

8 del Río, L.A., Sandalio, L.M., Corpas, F.J., Palma, J.M., and Barroso, J.B. (2006) *Plant Physiol.*, **141**, 330–335.

9 Shapiro, A.D. (2005) *Vitam. Horm.*, **72**, 339–398.

10 Besson-Bard, A., Pugin, A., and Wendehenne, D. (2008) *Annu Rev. Plant Biol.*, **59**, 21–39.

11 Corpas, F.J., Barroso, J.B., Carreras, A., Quirós, M., León, A.M., Romero-Puertas, M.C., Esteban, F.J., Valderrama, R., Palma, J.M., Sandalio, L.M., Gómez, M., and del Río, L.A. (2004) *Plant Physiol.*, **136**, 2722–2733.

12 Zemojtel, T., Frohlich, A., Palmieri, M.C., Kolanczyk, M., Mikula, I., Wyrwicz, L.S., Wanker, E.E., Mundlos, S., Vingron, M., Martasek, P., and Durner, J. (2006) *Trends Plant Sci.*, **11**, 524–525.

13 Crawford, N.M., Galli, M., Tischner, R., Heimer, Y.M., Okamoto, M., and Mack, A. (2006) *Trends Plant Sci.*, **11**, 526–527.

14 Pfeiffer, S., Mayer, B., and Hemmens, B. (1999) *Angew. Chem. Int. Ed.*, **38**, 1714–1731.

15 Gorren, A.C.F., Schrammel, A., Schmidt, K., and Mayer, B. (1996) *Arch. Biochem. Biophys.*, **330**, 219–228.

16 Halliwell, B. and Gutteridge, J.M.C. (2007) *Free Radicals in Biology and Medicine*, 4th edn, Oxford University Press, Oxford.

17 Stamler, J.S., Singel, D.J., and Loscalzo, J. (1992) *Science*, **258**, 1898–1902.

18 Cooper, C.E. (1999) *Biochim. Biophys. Acta*, **1411**, 290–309.

19 Thomas, D.D., Miranda, K.M., Colton, C.A., Citrin, D., Espey, M.G., and Wink, D.A. (2003) *Antioxid. Redox Signal.*, **5**, 307–317.

20 Lindermayr, C., Saalbach, G., and Durner, J. (2005) *Plant Physiol.*, **137**, 921–930.

21 Romero-Puertas, M.C., Campostrini, N., Matt,è A., Righetti, P.G., Perazzolli, M., Zolla, L., Roepstorff, P., and Delledonne, M. (2008) *Proteomics*, **8**, 1459–1469.

22 Rubbo, H. and Radi, R. (2008) *Biochim. Biophys. Acta*, **1780**, 1318–1324.

23 Gong, Y.W. and Yuan, Y.J. (2005) *J. Biotechnol.*, **123**, 185–192.

24 Morot-Gaudry-Talarmain, Y., Rockel, P., Moureaux, T., Quillere, I., Leydecker, M.T., Kaiser, W.M., and Morot-Gaudry, J.F. (2002) *Planta*, **215**, 708–715.

25 Saito, S., Yamamoto-Katou, A., Yoshioka, H., Doke, N., and Kawakita, K. (2006) *Plant Cell Physiol.*, **47**, 689–697.

26 Ischiropoulos, H. (2003) *Biochem. Biophys. Res. Commun.*, **305**, 776–783.

27 Freeman, B.A., Baker, P.R., Schopfer, F.J., Woodcock, S.R., Napolitano, A., and d'Ischia, M. (2008) *J. Biol. Chem.*, **283**, 15515–15519.

28 Trostchansky, A. and Rubbo, H. (2008) *Free Radic. Biol. Med.*, **44**, 1887–1896.

29 Cui, T., Schopfer, F.J., Zhang, J., Chen, K., Ichikawa, T., Baker, P.R., Batthyany, C., Chacko, B.K., Feng, X., Patel, R.P., Agarwal, A., Freeman, B.A., and Chen, Y. E. (2006) *J. Biol. Chem.*, **281**, 35686–35698.

30 Ohshima, H., Yoshie, Y., Auriol, S., and Gilibert, I. (1998) *Free Radic. Biol. Med.*, **25**, 1057–1065.

31 Terasaki, Y., Akuta, T., Terasaki, M., Sawa, T., Mori, T., Okamoto, T., Ozaki, M.,

Takeya, M., and Akaike, T. (2006) *Am. J. Respir. Crit. Care Med.*, **174**, 665–673.
32 Sawa, T., Tatemichi, T., Akaike, T., Barbina, A., and Ohshima, H. (2006) *Free Radic. Biol. Med.*, **40**, 711–720.
33 Polverari, A., Molesini, B., Pezzotti, M., Buonaurio, R., Marte, M., and Delledonne, M. (2003) *Mol. Plant–Microbe Interact.*, **16**, 1094–1105.
34 Parani, M., Rudrabhatla, S., Myers, R., Weirich, H., Smith, B., Leaman, D., and Goldman, S. (2004) *Plant Biotechnol. J.*, **2**, 359–366.
35 Bogdan, C. (2001) *Trends Cell Biol.*, **11** (2), 66–75.
36 Grun, S., Lindermayr, C., Sell, S., and Durner, J. (2006) *J. Exp. Bot.*, **57**, 507–516.
37 Hmadcha, A., Bedova, F., Sobrino, F., and Pintado, E. (1999) *J. Exp. Med.*, **190**, 1595–1604.
38 Hentze, M. and Kuhn, L. (1996) *Proc. Natl. Acad. Sci. USA*, **93**, 8175–8182.
39 Gardner, P.R., Costantino, G., Szabo, C., and Salzman, A.L. (1997) *J. Biol. Chem.*, **272**, 25071–25076.
40 Kim, S. and Ponka, P. (2002) *Proc. Natl. Acad. Sci. USA*, **99**, 12214–12219.
41 Kim, S., Wing, S.S., and Ponka, P. (2004) *Mol. Cell Biol.*, **24**, 330–337.
42 Arnaud, N., Murgia, I., Boucherez, J., Briat, J., Cellier, F., and Gaymard, F. (2006) *J. Biol. Chem.*, **281**, 23579–23588.
43 Matthews, J.R., Botting, C.H., Panico, M., Morris, H.R., and Hay, R.T. (1996) *Nucleic Acids Res.*, **24**, 2236–2242.
44 Khan, B.V., Harrison, D.G., Olbrych, M.T., Alexander, R.W., and Medford, R.M. (1996) *Proc. Natl. Acad. Sci. USA*, **93**, 9114–9119.
45 Stancoski, I. and Baltimore, D. (1997) *Cell*, **91**, 299–302.
46 Yakovlev, V.A., Barani, I.J., Rabender, C.S., Black, S.M., Leach, J.K., Graves, P.R., Kellogg, G.E., and Mikkelsen, R.B. (2007) *Biochemistry*, **46**, 11671–11683.
47 Hausladen, A., Privalle, C.T., Keng, T., DeAngelo, J., and Stamler, J.S. (1996) *Cell*, **86**, 719–729.
48 Klatt, P., Pineda-Molina, E., and Lamas, S. (1999) *J. Biol. Chem.*, **274**, 15857–15864.
49 Brendeford, E.M., Andersson, K.B., and Gabrielsen, O.S. (1998) *FEBS Lett.*, **425**, 52–56.
50 Serpa, V., Vernal, J., Lamattina, L., Grotewold, E., Cassia, R., and Terenzi, H. (2007) *Biochem. Biophys. Res. Commun.*, **361**, 1048–1053.
51 Cobbs, C.S., Whisenhunt, T.R., Wesemann, D.R., Harkins, L.E., Van Meir, E.G., and Samanta, M. (2003) *Cancer Res.*, **63**, 8670–8867.
52 Ding, H. and Demple, B. (2000) *Proc. Natl. Acad. Sci. USA*, **97**, 5146–5150.
53 Shinyashiki, M., Chiang, K.T., Switzer, C.H., Gralla, E.B., Valentine, J.S., Thiele, D.J., and Fukuto, J.M. (2000) *Proc. Natl. Acad. Sci. USA*, **97**, 2491–2496.
54 Kröncke, K.D. and Carlberg, C. (2000) *FASEB J.*, **14**, 166–173.
55 Chiranand, W., McLeod, I., Zhou, H., Lynn, J.J., Vega, L.A., Myers, H., Yates, J.R., Lorenz, M.C., and Gustin, M.C. (2008) *Eukaryot. Cell*, **7**, 268–278.
56 Palmieri, M.C., Sell, S., Huang, X., Scherf, M., Werner, T., Durner, J., and Lindermayr, C. (2008) *J. Exp. Bot.*, **59**, 177–186.
57 Valderrama, R., Corpas, F.J., Carreras, A., Fernández-Ocaña, A., Chaki, M., Luque, F., Gómez-Rodríguez, M.V., Colmenero-Varea, P., del Río, L.A., and Barroso, J.B. (2007) *FEBS Lett.*, **581**, 453–461.
58 Hernández, J.A., Corpas, F.J., Gómez, M., del Río, L.A., and Sevilla, F. (1993) *Physiol. Plant.*, **89**, 103–110.
59 Hernández, J.A., Olmos, E., Corpas, F.J., Sevilla, F., and del Río, L.A. (1995) *Plant Sci.*, **105**, 151–167.
60 Corpas, F.J., Gómez, M., Hernández, J.A., and del Río, L.A. (1993) *J. Plant. Physiol.*, **141**, 160–165.
61 Valderrama, R., Corpas, F.J., Carreras, A., Gómez-Rodríguez, M.V., Chaki, M., Pedrajas, J.R., Fernández-Ocaña, A., del Río, L.A., and Barroso, J.B. (2006) *Plant Cell Environ.*, **29**, 1449–1459.

62 Zhang, Y., Wang, L., Liu, Y., Zhang, Q., Wei, Q., and Zhang, W. (2006) *Planta*, **224**, 545–555.

63 Kopyra, M. and Gwózdz, E. (2003) *Plant Physiol. Biochem.*, **41**, 1011–1017.

64 Zhao, M.G., Tian, Q.Y., and Zhang, W.H. (2007) *Plant Physiol.*, **144**, 206–217.

65 Mackerness, S.A.H., John, C.F., Jordan, B., and Thomas, B. (2001) *FEBS Lett.*, **489**, 237–242.

66 Shi, S., Wang, G., Wang, Y., Zhang, L., and Zhang, L. (2005) *Nitric Oxide*, **13**, 1–9.

67 He, J.-M., Xu, H., She, X.-P., Song, X.-G., and Zhao, W.-M. (2005) *Funct. Plant Biol.*, **32**, 237–247.

68 Bright, J., Desikan, R., Hancock, J.T., Weir, I.S., and Neill, S.J. (2006) *Plant J.*, **45**, 113–122.

69 Rao, M.V. and Davis, K.R. (2001) *Planta*, **213**, 682–690.

70 Ederli, L., Morettini, R., Borgogni, A., Wasternack, C., Miersch, O., Reale, L., Ferranti, F., Tosti, N., and Pasqualini, S. (2006) *Plant Physiol.*, **142**, 595–608.

71 Velikova, V., Pinelli, P., Pasqualini, S., Reale, L., Ferranti, F., and Loreto, F. (2005) *New Phytol.*, **166**, 419–425.

72 Reymond, P., Weber, H., Damond, M., and Farmer, E.E. (2000) *Plant Cell*, **12**, 707–720.

73 Schilmiller, A.L. and Howe, G.A. (2005) *Curr. Opin. Plant Biol.*, **8**, 369–377.

74 Orozco-Cárdenas, M.L. and Ryan, C.A. (2002) *Plant Physiol.*, **130**, 487–493.

75 Huang, X., Stettmaier, K., Michel, C., Hutzler, P., Mueller, M.J., and Durner, J. (2004) *Planta*, **218**, 938–946.

76 Corpas, F.J., Chaki, M., Fernández-Ocaña, A., Valderrama, R., Palma, J.M., Carreras, A., Begara-Morales, J.C., Airaki, M., del Río, L.A., and Barroso, J.B. (2008) *Plant Cell Physiol.* doi: 10.1093/pcp/pcn144

77 Paris, R., Lamattina, L., and Casalongué, C.A. (2007) *Plant Physiol. Biochem.*, **45**, 80–86.

78 Loganathan, P., Hedley, M.J., and Grace, N.D. (2008) *Rev. Environ. Cont. Tox.*, **192**, 29–66.

79 Wagner, G.J. (1993) *Adv. Agron.*, **51**, 173–212.

80 Dixit, V., Pandey, V., and Shyam, R. (2001) *J. Exp. Bot.*, **52**, 1101–1109.

81 Romero-Puertas, M.C., Rodríguez-Serrano, M., Corpas, F.J., Gómez, M., del Río, L.A., and Sandalio, L.M. (2004) *Plant Cell Environ.*, **27**, 1122–1134.

82 Rodríguez-Serrano, M., Romero-Puertas, M.C., Zabalza, A., Corpas, F.J., Gómez, M., del Río, L.A., and Sandalio, L.M. (2006) *Plant Cell Environ.*, **29**, 1532–1544.

83 Barroso, J.B., Corpas, F.J., Carreras, A., Rodríguez-Serrano, M., Esteban, F.J., Fernández-Ocaña, A., Chaki, M., Romero-Puertas, M.C., Valderrama, R., Sandalio, L.M., and del Río, L.A. (2006) *J. Exp. Bot.*, **57**, 1785–1793.

84 Noriega, G.O., Yannarelli, G.G., Balestrasse, K.B., Batlle, A., and Tomaro, M.L. (2007) *Planta*, **226**, 1155–1163.

85 Laspina, V.N., Groppas, M.D., Tomaro, M.L., and Benavides, M.P. (2005) *Plant Sci.*, **169**, 323–330.

86 Delhaize, E. and Ryan, P.R. (1995) *Plant Physiol.*, **107**, 315–321.

87 Ma, J.F., Ryan, P.R., and Delhaize, E. (2001) *Trends Plant Sci.*, **6**, 273–278.

88 Tian, Q.Y., Sun, D.H., Zhao, M.G., and Zhang, W.H. (2007) *New Phytol.*, **174**, 322–331.

89 Wang, Y.S. and Yang, Z.M. (2005) *Plant Cell Physiol.*, **46**, 1915–1923.

90 Illés, P., Schlicht, M., Pavlovkin, J., Lichtscheidl, I., Baluska, F., and Ovecka, M. (2006) *J. Exp. Bot.*, **57**, 4201–4213.

5
Polyamines and Cytokinin: Is Nitric Oxide Biosynthesis the Key to Overlapping Functions?

Rinukshi Wimalasekera and Günther F.E. Scherer

Summary

The number of signals that are known to induce NO biosynthesis is steadily increasing. We have found that polyamines and cytokinin stimulate NO biosynthesis but with different kinetics. Polyamines stimulate NO biosynthesis without a lag phase while cytokinin stimulates NO biosynthesis with a brief (3 min) lag phase. When the tissue distribution pattern of NO is compared by fluorescence microscopy, polyamine- and cytokinin-treated *Arabidopsis* seedlings show a high degree of similarity. Highest NO-dependent fluorescence was found in the shoot meristem, young leaves, vascular bundles, trichomes, and root meristem. Moreover, there are a number of similarities in cytokinin- and polyamine-mediated physiological functions such as embryogenesis, flowering, and senescence, and in plant response to abiotic and biotic stresses, indicating overlapping functions of both signaling substances. Our goal is to determine the sources of NO biosynthesis to gain an understanding of the overlapping functions of polyamines and cytokinins.

5.1
Introduction

Increasing evidence indicates that nitric oxide (NO) is a key signaling molecule involved in a wide range of functions in plants. These functions include seed germination [1], inhibition of senescence [2], inhibition of fruit ripening [3], flowering [4], root development [5, 6], plant response to abiotic stresses, such as drought, heat, and chilling stress [7–9], and biotic stresses [7–19]. NO is also known as a signaling molecule in cytokinin [20–23], abscisic acid [24–28], auxin [5, 9, 29], ethylene [3], and polyamine (PA) [30] signaling. Unlike in animals, the mechanisms of NO biosynthesis in plants are not completely known. Several sources of NO may be

Nitric Oxide in Plant Physiology. Edited by S. Hayat, M. Mori, J. Pichtel, and A. Ahmad

ISBN: 978-3-527-32519-1

present in plants; for example, nitrate reductase (NR) [31, 32], a putative nitric oxide synthase (NOS) [33, 34], and nonenzymatic NO sources are described [7, 8, 34].

Cytokinins have been implicated in many aspects of plant growth and development including seed germination, cell proliferation, shoot organogenesis, regulation of stem and root growth, apical dominance, chloroplast differentiation, prevention of leaf senescence, and anthocyanine accumulation [35, 36]. Recently, it has been shown that NO synthesis was induced by cytokinin [20–23]. In *Amaranthus* seedlings, cytokinin action such as betalaine accumulation was shown to be mimicked by donors of NO and inhibited by arginine-related NO synthase (NOS) inhibitors [20].

Polyamines are low molecular weight, polycationic, nitrogenous compounds present in all living cells that are implicated in various developmental processes. PAs often exist not only as free molecular bases but also as conjugated and bound forms [37–39]. The most common PAs in plants are diamine putrescine, triamine spermidine, and tetraamine spermine. In plants, PAs play a role in physiological processes such as regulation of cell proliferation and differentiation, dormancy breaking of tubers and seed germination, development of flowers and fruits, zygotic and somatic embryogenesis, and senescence [38–41]. PAs are known to enhance tolerance to environmental stresses such as salinity, chilling, drought, and potassium deficiency [38, 42, 43], and defense signaling against pathogens [44]. In a recent study, it has been shown that PAs induced NO biosynthesis in *Arabidopsis* seedlings, giving a new insight into PA-mediated signaling and NO as a potential mediator of PA actions [30].

Taken together, the biology of polyamines, cytokinins, and NO seems to possess many overlapping functions. These issues inspired us to study the relationship between PAs and NO and the relationship between cytokinins and NO.

5.2
Cytokinin- and Polyamine-Induced NO Biosynthesis

Recent experimental evidence in cytokinin- and PA-induced NO biosynthesis suggests NO as a signaling intermediate in cytokinin and PA action.

When tobacco BY-2 cells and parsley and *Arabidopsis* cell cultures were treated with 0–6 μM of zeatin and benzyl aminopurine (BAP), an increase in NO release was observed above endogenous NO levels as measured by fluorometric assay using DAF-2. The NO release was rapid and apparent after 3 min [21]. The effect of different concentrations of BAP and duration of treatment on NO release was demonstrated in *Arabidopsis* cells; a 34% increase in DAF-2-mediated fluorescence above the endogenous level was observed with 27 μM BAP treatment [22]. Moreover, enhancement of NO release was observed in 0–20 μM zeatin-treated 7-day-old *Arabidopsis* seedlings, and the NO release was dependent on zeatin concentration [23].

Treatment of cell cultures and the *Arabidopsis* seedlings with NO scavenger 2-(4-carboxyphenyl)-4,4,5,5-tetra-methylimidazoline-1-1-oxy-3-oxide (PTIO) inhibited the BAP/zeatin-induced NO accumulation, verifying that the effects were NO-specific

[21–23]. Zeatin- and BAP-induced NO synthesis was quenched by an inhibitor of animal NO synthase, 2-(aminoethyl)-2-thiopseudourea (AET) suggesting that a NOS-like activity is a potential contributor of NO synthesis [21–23]. Endogenous NO release from *nia1,2 Arabidopsis* seedlings that were deficient in nitrate reductase was rapidly enhanced by zeatin treatment in a concentration-dependent manner similar to wild type, indicating that some other enzyme(s), not NR, is/are the zeatin-regulated NO producing enzyme(s) [23]. However, *nia1,2* seedlings showed that NR is essential for long-term regulation of NO balance in plants because its absence caused an aberrant response to zeatin [23]. In *nia1,2* seedlings, zeatin-induced NO synthesis was inhibited by AET, suggesting the relevant enzyme was AET-sensitive.

A fluorometric assay using cell-impermeable dye DAR-4M showed that NO production was significantly induced above endogenous levels in *Arabidopsis* seedlings as well as in tobacco BY-2 cells when treated with 1 mM of PAs spermine, spermidine, and putrescine, with spermine being the most active and putrescine a weak PA [30]. A very rapid increase in NO release without an apparent lag phase was observed in 1 mM spermine-treated seedlings [30]. Both endogenous and PA-induced NO induction was inhibited by PTIO and AET [30]. In *nia1,2* plants, PA-induced fluorescence in the medium was observed, suggesting NR is not involved in NO production (unpublished results). NOA (NO-associated protein) as an enzymatic source remains unclear. It can be speculated that yet unknown enzyme/s or PA oxidases are responsible for NO synthesis [30, 45]. Rapid induction of NO release by spermine without an apparent lag phase favors the presence of an enzyme that directly converts PA to NO.

It is hypothesized that PA- and cytokinin-regulated NO biosynthesis implies the contribution of similar or identical enzymes, but their identity remains to be found.

5.3
Tissue Distribution of Zeatin-Induced and PA-Induced NO Formation

When *Arabidopsis* seedlings were treated with 5 μM zeatin, strong increases in NO-dependent fluorescence as determined using cell-permeable DAR-4M-AM were observed in leaves, especially in trichomes, hydathodes, cuticles of guard cells, cotyledons, root–shoot transition zones, and elongation zones of root tips [23]. A clear decrease in fluorescence was seen when zeatin-treated seedlings were treated with PTIO, indicating fluorescence was caused by NO accumulation [23]. The tissue pattern of zeatin-induced fluorescence in *Atnoa1*, which lacks putative AtNOA1, was comparable to wild type; likewise, the physiological response was unchanged in *Atnoa1* in response to zeatin, suggesting AtNOA1 is not regulated by zeatin in NO induction [23]. Mutant *nia1,2*, lacking NR, responded to zeatin by enhanced fluorescence formation in leaves, cotyledons, root–shoot transition zones, and elongation zones of root tips [23]. However, in *nia1,2*, stronger fluorescence in cotyledons than in wild type was observed in response to zeatin, and was highest at 0.5 μM zeatin. In comparison to wild type, *nia1,2* showed hypersensitivity to zeatin, producing larger cotyledons, delay in expansion of leaves, and thicker and longer hypocotyls [23]. These observations suggest the absence of NR as a NO source alters the response to zeatin, thus identifying NO as a mediator of cytokinin action. The altered physiological responses observed in *nia1,2* may also be linked to mutual interplay of cytokinin, NR, and nitrogen metabolism. It has been shown that increase in nitrogen sources upregulates elements of cytokinin signal transduction [46, 47].

When *Arabidopsis* seedlings were treated with PAs spermine, spermidine, and putrescine, an increase in tissue-specific distribution of NO-dependent fluorescence, which was very similar to the zeatin-induced NO distribution pattern, was observed [30]. Spermine and spermidine induced higher NO-dependent fluorescence in the elongation zone of the root tip and primary leaves compared to endogenous levels. In cotyledons, only a weak increase or no increase in fluorescence was found. As with zeatin-induced NO-dependent fluorescence, PTIO quenched PA-induced fluorescence increase in the leaves and roots suggesting true NO biosynthesis. Since the enzymatic source of PA-induced NO biosynthesis is unclear, an explanation for the tissue distribution pattern could only be speculative [30].

5.4 Nitric Oxide, Cytokinin, and Polyamines in Plant Growth and Development and in Abiotic and Biotic Stresses

Cytokinin is an essential phytohormone involved in almost all aspects of plant growth and development [36]. Polyamines are active at various concentrations ranging from micromolar to millimolar. In spite of their higher concentrations in the plant compared to those of phytohormones, PAs are reported to mediate the action of hormones and are known to play a key role in plant growth and development and plant defense. Since the discovery of NO in plants, it has become increasingly evident that NO is an essential molecule that mediates diverse growth and development processes as well as defense responses. PA- and cytokinin-induced NO biosynthesis might be a potential linking signal to many of these functions in plants.

5.4.1 Embryogenesis

Cytokinins are known to play many essential roles in postembryonic growth and development, but their role in early embryogenesis remains doubtful [48, 49]. Cytokinin and auxin interaction is known to be critical for early embryogenesis signaling [50]. Although a direct effect of cytokinin-induced NO on embryogenesis has not been reported, NO may serve as an intermediate signaling molecule.

PAs are crucial components in embryogenesis [38, 41, 42, 51]. Exogenous application of PAs to suspension cultures of *Ocotea catharinensis*, *Araucaria angustifolia*, and *Pinus taeda* were shown to alter the endogenous NO in somatic embryos, suggesting PA-induced NO may play an important role in embryogenesis [52–56].

5.4.2
Flowering

Implications of cytokinin for floral initiation have been documented. It has been demonstrated that an increase in endogenous cytokinin content occurs in the shoot apex at floral transition [57]. Mutants that overproduce or are deficient in cytokinin showed altered flowering times [58]. Endogenous PA concentration in plants is known to change dramatically during floral transition, and delay in flowering is correlated with a reduction in polyamine accumulation [42, 59, 60]. NO is known to repress floral transition in *Arabidopsis* [4]. Plants treated with NO and a mutant *nox1* that overproduce NO showed delayed flowering whereas *nos1* plants producing less NO showed early flowering [4]. Since cytokinin, PA, and NO are involved in the regulation of floral transition, they may share some common signal transduction pathway.

5.4.3
Senescence

Another common character of NO, PA, and cytokinin is the antisenescence property. Ethylene plays an active role in senescence processes in plants. Recent evidence with several plant species such as pea, rice, and many varieties of flowers and fruits indicates that NO possesses antisenescence properties [2, 7]. NO and ethylene impart antagonistic effects on senescence of plants. Exogenous application of NO extends the postharvest life of horticultural products by inhibiting ethylene production [2]. In addition to the inhibition of ethylene formation, the NO-mediated antisenescence effect may occur by counteracting oxidative stress and scavenging free radicals. Fruit ripening is another senescence-related process promoted by ethylene that can be retarded by NO [7]. The antisenescence effects of PAs have been reported in *Arabidopsis*, rose, and rice leaves [61–63]. These effects may occur via retention of chlorophyll, inhibition of ethylene synthesis, stabilization of membranes, prevention of lipid peroxidation, and scavenging of free radicals [61–63]. Spermine and spermidine were shown to reduce ethylene synthesis by inhibition of ethylene precursor ACC [38]. Cytokinins are well known for their antisenescence properties [36, 64]. Treating detached leaves with cytokinin delays their senescence, suggesting that cytokinins are natural regulators of leaf senescence. This effect of cytokinin-stimulated and PA-stimulated NO biosynthesis on *Arabidopsis* has recently been shown [21, 23, 30]. Taken together, it can be proposed that there may be a link among NO, cytokinin, and PA in antisenescence effects.

5.4.4
Programmed Cell Death

An association is recognized between NO and PCD in different plant systems, either induced by pathogens or by mechanical stress [7, 65, 66]. Increasing evidence indicates an interaction between ROS and NO in the regulation of PCD [7]. Further, involvement of a NOS-like enzyme and inhibition of respiration in NO-induced PCD is described [7, 67]. Cytokinin-mediated PCD is evident in plants [22, 68, 69]. In a recent study, NO signaling was demonstrated in cytokinin-induced PCD. Cytokinin-induced PCD was reduced when cytokinin was supplemented with NO scavenger and NOS inhibitors [22]. PAs are known to be involved in cell death process called apoptosis in animal systems. Both upregulation and downregulation of polyamine levels have been reported during apoptosis [70]. Recently, in tobacco plants, it has been shown that PAs play a critical role in hypersensitive cell death in pathogen attack and are thought to be mediated by hydrogen peroxide, which is a catabolic product of PA by polyamine oxidases (PAOs) [71]. Evidence of association of NO with mediating PA-induced PCD is as yet lacking.

5.4.5
Abiotic Stresses

It is widely known that NO as well as PAs are associated with various abiotic stresses such as temperature extremes, drought, salinity, and mineral deficiency [7–9, 38, 42]. A prominent theme where PAs and NO appear to have overlapping functions is in defense responses against these stress conditions.

In heat-tolerant cotton and rice, substantial increases in free and conjugated PAs and long-chained PA, as well as greater accumulation of polyamine oxidases and PA-biosynthesizing arginine decarboxylase (ADC), were observed during heat stress [72, 73]. Similarly, differential accumulation of PAs in chill-tolerant wheat, zucchini, and rice has been demonstrated [74]. An increase in NO production was observed in alfalfa during short-term heat stress conditions, and exogenous NO was shown to mediate chilling resistance in tomato, wheat, and corn [75]. NO generated during heat and chilling conditions might be partly due to accumulated PAs. Cytokinin may play a role in coping with heat stress in plants. In creeping bentgrass and in a dwarf wheat variety, exposure to high temperature induced a decrease in cytokinin levels [76].

Increase in endogenous levels of PAs in response to salt stress has been reported in a number of plant species including rice, sorghum, and tomato, thus suggesting their function in salinity adaptive responses [77–79]. In most studies, concerning salt stress, the induced PA response has been assumed to rely primarily on ADC activation and stimulation of amine oxidase activities [80, 81]. It is speculated that PA-induced NO generation, possibly through amine oxidase activity, might be an intermediate candidate involved in salt stress tolerance. It has been demonstrated that NO enhances salt tolerance in maize [82]. In another study, sensitivity to NaCl during germination was observed in *Arabidopsis Atnoa1* plants with an impaired capacity to synthesize NO *in vivo*, suggesting NO role in salinity tolerance [83].

Increased NO biosynthesis during osmotic stress has been reported in several plant species [7, 84–87]. Close interactions among ABA, ROS, and NO in dealing with osmotic stress have been suggested. In response to drought stress, an increase in NOS-like activity was observed in wheat seedlings, and ABA accumulation was inhibited by NOS inhibitors [88]. An accumulation of putrescine levels is a common observation in plants under osmotic stress [38]. Eventual synthesis of other PA from putrescine is the key protective factor for the stressed cells. Regulation of ADC biosynthesis and polyamine oxidases during osmotic stress was described [38]. PA-dependent NO production might be carried out by PAO or an unknown enzyme. Physiological studies with PAO mutants may provide a clue to their participation in NO synthesis during osmotic stress.

During drought, a significant increase in cytokinin levels was observed in leaves of tobacco plants as revealed by expression of IPTgene, driven by maturation- and stress-induced promoters. These data suggest that increased production of cytokinin contributed to enhanced drought tolerance [89]. However, the mechanism of enhanced cytokinin leading to activation of the drought tolerance mechanism is not yet identified.

Accumulation of putrescine in leaves of K^+-deficient plants is a commonly observed phenomenon [38]. The degree of putrescine accumulation is correlated with the appearance of severe K^+ deficiency symptoms [90]. Changes in PA composition and content induced by K^+-deficient conditions are associated with alterations in plant developmental stages such as delayed flowering. In addition, Mg^{2+} deficiency is known to induce PA accumulation. A function of NO in K^+ deficiency has not yet been documented, however. Recently, it was found that NO has a function in response to iron deficiency [91].

When all the above-mentioned abiotic stresses are taken into consideration, PA and NO appear to play a role in certain response actions. NO may be a link between PA-mediated stress responses. It is not known whether catabolism of PA by polyamine oxidases generates NO or not; this remains an open question.

5.4.6 Biotic Stresses

There is substantial evidence for NO in exerting defense responses in various plant–pathogen interactions. NO is involved in defense responses such as hypersensitive response (HR), reactive oxygen species (ROS) burst, and induction of pathogenesis-related gene such as *PR-1*, *PAL*, and salicylic acid [7–19, 92–94]. The ROS burst is very rapid upon signal application and the overall time course of ROS and NO generation occurs within a few minutes [95].

Several roles of PAs and their catabolism in plant pathogen defense have been extensively demonstrated. ROS produced from PA trigger the HR and PCD. Spermine acts as an inducer of PR proteins and as a trigger for caspase activity leading to HR. Accumulation of free spermine in intercellular spaces of tobacco leaves exhibiting HR to tobacco mosaic virus infection was observed. Spermine caused mitochondrial dysfunction in a signaling pathway with the upregulation of protein kinases (PKs) such as WIPK (wound-induced protein kinase), SIPK (salicylic acid-induced protein kinase), and MAPK (mitogen-activated protein kinase) [96–98]. NO is known to activate protein kinases in tobacco and *Arabidopsis* in pathogen-induced signaling [11, 12]. The rapid NO increase with addition of PA found in our experiments suggests that NO could be the upstream component of PA-induced

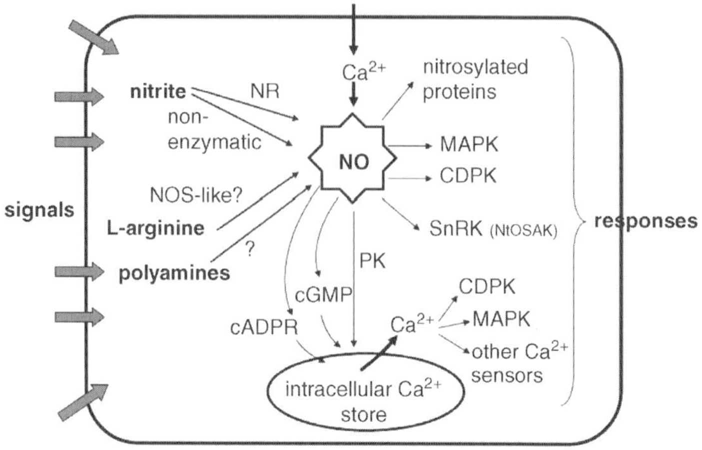

Figure 5.1 Schematic representation of NO signaling in plant cells. When plant cells are exposed to biotic and abiotic signals, NO can be synthesized from nitrate, catalyzed by nitrate reductase and a nonenzymatic pathway, from L-arginine that involves an uncharacterized nitric oxide synthase-like enzyme (NOS-like) and through an unidentified enzyme that could use polyamine as a substrate. NO induces increase in cytosolic Ca^{2+} by extracellular Ca^{2+} uptake and by activation of intracellular Ca^{2+} channels. cyclic ADP ribose (cADPR), cyclic GMP, and protein kinases may be involved in this mechanism. The activity of CDPKs, MAPKs, and NtOSAK, a member of sucrose nonfermenting 1-related protein kinase 2, are regulated by increased cytosolic Ca^{2+} concentrations. NO modulates some cellular responses through protein nitrosylation. The interplay among these targets leads to diverse NO-triggered responses such as defense gene expression and stomatal closure. (Modified from Refs [33, 34].)

MAPK in defense responses. Regarding the induction of defense-related genes, PAs and NO appear to have similar functions and NO induction by PAs may be a prior element of PA-induced defense gene expression.

Numerous cellular responses mediated by NO are regulated by an array of signaling components such as Ca^{2+}, protein kinases, Ca^{2+}-dependent protein kinases (CDPKs), mitogen-activated protein kinases, sucrose nonfermenting 1-related protein kinase 2 (SnRK2), cyclic GMP (cGMP), and nitrosylated proteins (Figure 5.1) [33, 34].

Understanding how NO modulates the activities of these signaling components and their interactions toward specific cellular response/s in plants is still very limited. NO-stimulated regulation of Ca^{2+} mobilization is described in diverse responses such as osmotic stress tolerance, stomatal closure, defense gene expression, and adventitious root formation [11, 24, 33, 34, 92, 99, 100]. These investigations support possible links among NO, cGMP, cADPR, CDPK, and Ca^{2+}. NtOSAK, a protein kinase that belongs to the SnRK2 member was identified in *Nicotiana tabacum* subjected to osmotic stress [101]. The significance of protein nitrosylation in plant cell regulation is emerging [94, 102–105]. Specific combination of second messengers associated with NO signaling is a crucial aspect in determining a particular response.

The specificity of NO signaling is a key issue to be understood. In animal cells, the importance of temporal and spatial arrangement of NO signaling in response specificity is described [34, 106]. The specificity of NO signaling might be regulated by compartmental localization, kinetics, and intensity of NO production. Localization specificity of NO sources could be an important aspect in determining the specificity of responses toward a certain signal.

References

1 Beligni, M.V. and Lamattina, L. (2000) *Planta*, **210**, 215–221.

2 Leshem, Y.Y., Wills, R.B.H., and Veng-Va Ku, V. (1998) *Plant Physiol. Biochem.*, **36**, 825–833.

3 Leshem, Y.Y. and Pinchasov, Y. (2000) *J. Exp. Bot.*, **51**, 1471–1473.

4 He, Y., Tang, R.H., Hao, Y., Stevens, R.D., Cook, C.W., Ahn, S.M., Jing, L., Yang, Z., Chen, L., Guo, F., Fiorani, F., Jackson, R.B., Crawford, N.M., and Pei, Z.-M. (2004) *Science*, **305**, 1968–1971.

5 Pagnussat, G.C., Simontacchi, M., Puntarulo, S., and Lamattina, L. (2002) *Plant Physiol.*, **129**, 954–956.

6 Kolbert, Z., Bartha, B., and Erdei, L. (2008) *Physiol. Plant.*, **133**, 406–416.

7 Neill, S.J., Desikan, R., and Hancock, J.T. (2003) *New Phytologist.*, **159**, 11–35.

8 Neill, S., Barros, R., Bright, J., Desikan, R., Hancock, J., Harrison, J., Morris, P., Ribeiro, D., and Wilson, I. (2008) *J. Exp. Bot.*, **59**, 165–176.

9 Erdei, L. and Kolbert, Z. (2008) *Acta Biol. Szeg.*, **52**, 1–5.

10 Delledonne, M., Xia, Y., Dixon, R.A., and Lamb, C. (1998) *Nature*, **394**, 585–588.

11 Durner, J., Wendehenne, D., and Klessig, D.F. (1998) *Proc. Natl. Acad. Sci. USA*, **95**, 10328–10333.

12 Clarke, A., Desikan, R., Hurst, R.D., Hancock, J.T., and Neill, S.J. (2000) *Plant J.*, **24**, 667–677.

13 Delledonne, M., Zeier, J., Marocco, A., and Lamb, C. (2001) *Proc. Natl. Acad. Sci. USA*, **98**, 13454–13459.

14 Wendehenne, D., Pugin, A., Klessig, D.F., and Durner, J. (2001) *Trends Plant Sci.*, **6**, 177–183.

15 Huang, X., Kiefer, E., Rad, U.V., Ernst, D., Foissner, I., and Durner, J. (2002) *Plant Physiol. Biochem.*, **40**, 625–631.

16 Wendehenne, D., Durner, J., and Klessig, D.F. (2004) *Curr. Opin. Plant Biol.*, **7**, 449–455.

17 Romero-Puertas, M.C., Perazzolli, M., Zago, E.D., and Delledonne, M. (2004) *Cell. Microbiol.*, **6**, 795–803.

18 Delledonne, M. (2005) *Curr. Opin. Plant Biol.*, **8**, 390–396.

19 Mur, L.A.J., Carver, T.L.W., and Prats, E. (2006) *J. Exp. Bot.*, **57**, 489–505.

20 Scherer, G.F.E. and Holk, A. (2000) *Plant Growth Regul.*, **32**, 345–350.

21 Tun, N.N., Holk, A., and Scherer, G.F.E. (2001) *FEBS Lett.*, **509**, 174–176.

22 Carimi, F., Zottini, M., Costa, A., Cattalani, I., De Michele, M., Terzi, M., and Lo Schiavo, F. (2005) *Plant Cell Environ.*, **28**, 1171–1178.

23 Tun, N.N., Livaya, M., Kieber, J.J., and Scherer, G.F.E. (2008) *New Phytologist.*, **178**, 515–531.

24 Desikan, R., Griffiths, R., Hancock, J., and Neill, S. (2002) *Proc. Natl. Acad. Sci. USA*, **99**, 16314–16318.

25 Garcia-Mata, C. and Lamattina, L. (2002) *Plant Physiol.*, **128**, 790–792.

26 Lamattina, L., Garcia-Mata, C., Graziano, M., and Pagnussat, G. (2003) *Annu. Rev. Plant Biol.*, **54**, 109–136.

27 Guo, F.-Q., Okomoto, M., and Crawford, N.M. (2003) *Science*, **302**, 100–103.

28 Bright, J., Desikan, R., Hancock, J.T., Weir, I.S., and Neill, S.J. (2006) *Plant J.*, **45**, 113–122.

29 Kolbert, Z., Bartha, B., and Erdei, L. (2008) *J. Plant Physiol.*, **165**, 967–975.

30 Tun, N.N., Santa-Catarina, C., Begum, T., Silveira, V., Handro, W., Floh, I.S., and Scherer, G.F.E. (2006) *Plant Cell Physiol.*, **47**, 346–354.

31 Harper, J.E. (1981) *Plant Physiol.*, **68**, 1488–1493.

32 Stöhr, C., Strube, F., Marx, G., Ullrich, W.R., and Rockel, P. (2001) *Planta*, **212**, 835–841.

33 Courtois, C., Besson, A., Bourque, S., Dobrowolska, G., Pugin, A., and Wendehenne, D. (2008) *J. Exp. Bot.*, **59**, 155–163.

34 Besson-Bard, A., Pugin, A., and Wendehenne, D. (2008) *Annu. Rev. Plant Biol.*, **59**, 21–39.

35 Mok, D.W.S. and Mok, M.C. (1994) *Cytokinins: Chemistry, Activity and Function*, CRC Press, Boca Raton, FL, pp. 155–166.

36 Taiz, L. and Zeiger, E. (2002) *Plant Physiology*, 3rd edn, Sinauer Associates, Inc., Sunderland, MA, pp. 502–509.

37 Davies, P.J. (1990) *Plant Hormones and Their Role in Plant Growth and Development*, Kluwer Academic Publishers, Dordrecht, The Netherlands, pp. 280–295.

38 Bouchereau, A., Aziz, A., Larher, F., and Martin-Tanguy, J. (1999) *Plant Sci.*, **140**, 103–125.

39 Bagni, N. and Tassoni, A. (2001) *Amino Acids*, **20**, 301–317.

40 Kakkar, R.K., Nagar, P.K., Ahuja, P.S., and Rai, V.K. (2000) *Biologia Plant.*, **43**, 1–11.

41 Bais, P.H. and Ravishankar, G.A. (2002) *Plant Cell Tissue Organ. Cult.*, **69**, 1–34.

42 Galston, A.W., Kaur-Sawhney, R., Altabella, T., and Tiburcio, A.F. (1997) *Bot. Acta*, **110**, 197–207.

43 Liu, J.-H., Kitashiba, H., Wang, J., Ban, Y., and Moriguchi, T. (2007) *Plant Biotechnol.*, **24**, 117–126.

44 Walters, D. (2003) *New Phytologist.*, **159**, 109–115.

45 Yamasaki, H. and Cohen, M.F. (2006) *Trends Plant Sci.*, **11**, 522–524.

46 Lu, J.L., Ertl, J.R., and Chen, C.M. (1990) *Plant Mol. Biol.*, **14**, 585–594.

47 Takei, K., Takahashi, T., Sugiyama, T., Yamaya, T., and Sakakibara, H. (2002) *J. Exp. Bot.*, **53**, 971–977.

48 Nishimura, C., Ohashi, Y., Sato, S., Kato, T., Tabata, S., and Ueguchi, C. (2004) *Plant Cell*, **16**, 1365–1377.

49 Riefler, M., Novak, O., Strnad, M., and Schmülling, T. (2006) *Plant Cell*, **18**, 40–54.

50 Müller, B. and Sheen, J. (2008) *Nature*, **453**, 1094–1098.

51 Bertoldi, D., Tassoni, A., Martinelli, L., and Bagni, N. (2004) *Physiol. Plant.*, **120**, 657–666.

52 Santa-Catarina, C., Randi, A.M., and Viana, A.M. (2003) *Plant Cell Tissue Organ. Cult.*, **74**, 67–71.

53 Santa-Catarina, C., Hanai, L.R., Dornelas, M.C., Viana, A.M., and Floh, E.I.S. (2004) *Plant Cell Tissue Organ. Cult.*, **79**, 53–61.

54 Silveira, V., Floh, E.I.S., Handro, W., and Guerra, M. (2004) *Plant Cell Tissue Organ. Cult.*, **76**, 53–61.

55 Silveira, V., Santa-Catarina, C., Tun, N.N., Scherer, G.F.E., Handro, W., Guerra, M.P., and Floh, E.I.S. (2006) *Plant Sci.*, **171**, 91–98.

56 Santa-Catarina, C., Silveira, V., Scherer, G.F.E., and Floh, E.I.S. (2007) *Plant Cell Tissue Organ. Cult*, **90**, 93–101.

57 Corbesier, L., Prinsen, L., Jacqmard, A., Lejeune, P., Van Onckelen, H., Périlleux, C., and Bernier, G. (2003) *J. Exp. Bot.*, **54**, 2511–2517.

58 Eckardt, N.A. (2003) *Plant Cell*, **15**, 2489–2492.

59 Kakkar, R.K. and Sawhney, V.K. (2002) *Physiol. Plant.*, **116**, 281–292.

60 Martin-Tanguy, J., Tepfer, D., Paynot, M., Burtin, D., Heisler, L., and Martin, C. (1990) *Plant Physiol.*, **92**, 912–918.

61 Cheng, S.H., Shyr, Y.Y., and Kao, C.H. (1984) *Bot. Bull. Acad. Sin.*, **25**, 191–196.

62 Sood, S. and Nagar, P.K. (2003) *Plant Growth Regul.*, **39**, 155–160.

63 Woo, H.R., Kim, J.H., Nam, H.G., and Lim, P.O. (2004) *Plant Cell Physiol.*, **45**, 923–932.

64 Hajouj, T., Michelis, R., and Gepstein, S. (2000) *Plant Physiol.*, **124**, 1305–1314.

65 Clark, D., Durner, J., Navarre, D.A., and Klessig, D.F. (2000) *Mol. Plant–Microbe Interact.*, **13**, 1380–1384.

66 Pedroso, M.C., Magalhaes, J.R., and Durzan, D. (2000) *J. Exp. Bot.*, **51**, 1027–1036.

67 Zottini, M., Formentin, E., Scattolin, M., Carimi, F., Lo Schiavo, F., and Terzi, M. (2002) *FEBS Lett.*, **515**, 75–78.

68 Mlejnek, P. and Prochazka, S. (2002) *Planta*, **215**, 158–166.

69 Carimi, F., Zottini, M., Formentin, E., Terzi, M., and Lo Schiavo, F. (2003) *Planta*, **216**, 413–421.

70 Pignatti, C., Tantini, B., Stefanelli, C., and Flamigni, F. (2004) *Amino Acids*, **27**, 359–365.

71 Yoda, H., Yamaguchi, Y., and Sano, H. (2003) *Plant Physiol.*, **132**, 1973–1981.

72 Kuehn, G.D., Rodriguez-Garay, B., Bagga, S., and Phillips, G.C. (1990) *Plant Physiol.*, **94**, 855–857.

73 Roy, M. and Ghosh, B. (1996) *Physiol. Plant.*, **98**, 196–200.

74 Flores, H.E. (1991) Changes in polyamine metabolism in response to abiotic stress, in *Biochemistry and Physiology of Polyamines in Plants* (eds R.D. Slocum and H.E. Flores), CRC Press, Boca Raton, FL, pp. 213–225.

75 Lamattina, L., Beligni, M.V., Gracia-Mata, C., and Laxalt, A.M. (2001) US Patent 6242384 B1.

76 Wahid, A., Gelani, S., Ashraf, M., and Foolad, M.R. (2007) *Environ. Exp. Bot.*, **61**, 199–223.

77 Prakash, L. and Prathapasenan, G. (1988) *J. Agron. Crop Sci.*, **160**, 325–334.

78 Krishnamurthy, R. and Bhagwat, K.A. (1989) *Plant Physiol.*, **91**, 500–504.

79 Santa-Cruz, A., Perez-Alfocea, F., and Bolarin, M.C. (1997) *Physiol. Plant.*, **101**, 341–346.

80 Flores, H.E. and Galston, A.W. (1984) *Plant Physiol.*, **75**, 102–109.

81 Smith, T.A. (1985) *Biochem. Soc. Trans.*, **13**, 319–322.

82 Zhang, Y., Wang, L., Liu, Y., Zhang, Q., Wei, Q., and Zhang, W. (2006) *Planta*, **224**, 545–555.

83 Zhao, M.-G., Tian, Q.-Y., and Zhang, W.-H. (2007) *Plant Physiol.*, **144**, 206–217.

84 Leshem, Y.Y. and Haramaty, E. (1996) *J. Plant Physiol.*, **148**, 258–263.

85 Erdei, L., Szegeltes, Z., Barabas, K., and Pestenacz, A. (1996) *J. Plant Physiol.*, **147**, 599–603.

86 Neill, S.J., Desikan, R., Clarke, A., and Hancock, J.T. (2002) *Plant Physiol.*, **128**, 13–16.

87 Neill, S.J., Desikan, R., Clarke, A., Hurst, R.D., and Hancock, T. (2002) *J. Exp. Bot.*, **53**, 1237–1247.

88 Zhao, Z., Chen, G., and Zhang, C. (2001) *Aust. J. Plant Physiol.*, **28**, 1055–1061.

89 Rivero, R.M., Kojima, M., Gepstein, A., Sakakibara, H., Mittler, R., Gepstein, S., and Blumwald, E. (2007) *Proc. Natl. Acad. Sci. USA*, **104**, 19631–19636.

90 Geny, L., Broquedis, M., Martin-Tanguy, J., Soyer, J.P., and Bouard, J. (1997) *Am. J. Enol. Vitic.*, **48**, 85–92.

91 Murgia, I., Delledonne, M., and Soave, C. (2002) *Plant J.*, **30**, 521–528.

92 Lamotte, O., Gould, K., Lecourieux, D., Sequeira-Legrand, A., Lebrun-Garcia, A., Durner, J., Pugin, A., and Wendehenne, D. (2004) *Plant Physiol.*, **135**, 516–529.

93 Zeier, J., Delledonne, M., Mishina, T., Severi, E., Sonoda, M., and Lamb, C. (2004) *Plant Physiol.*, **136**, 2875–2886.

94 Romero-Puertas, M.C., Campostrini, N., Matte, A., Righetti, P.G., Perazzolli, M., Zolla, L., Roepstorff, P., and Delledonne, M. (2008) *Proteomics*, **8**, 1459–1469.

95 Foissner, I., Wendehenne, D., Langebartels, C., and Durner, J. (2000) *Plant J.*, **23**, 817–824.

96 Yamakawa, H., Kamada, H., Satoh, M., and Ohashi, Y. (1998) *Plant Physiol.*, **118**, 1213–1222.

97 Hiraga, S., Ito, H., Yamakawa, H., Ohtsubo, N., Seo, S., Mitsuhara, I., Matsui, H., Honma, M., and Ohashi, Y. (2000) *Mol. Plant–Microbe Interact.*, **13**, 210–216.

98 Takahashi, Y., Berberich, T., Miyazaki, A., Seo, S., Ohashi, Y., and Kusano, T. (2003) *Plant J.*, **36**, 820–829.

99 Garcia-Mata, C., Gay, R., Sokolovski, S., Hills, A., Lamattina, L., and Blatt, M.R. (2003) *Proc. Natl. Acad. Sci. USA*, **100**, 11116–11121.

100 Lanteri, M., Pagnussat, G.C., and Lamattina, L. (2006) *J. Exp. Bot.*, **57**, 1341–1351.

101 Kelner, A., Pekala, I., Kaczanowski, S., Muszynska, G., Hardie, D.G., and Dobrowolska, G. (2004) *Plant Physiol.*, **136**, 3255–3265.

102 Lindermayr, C., Saalbach, G., and Durner, J. (2005) *J. Plant Physiol.*, **137**, 921–930.

103 Lindermayr, C., Saalbach, G., Bahnweg, G., and Durner, J. (2006) *J. Biol. Chem.*, **281**, 4285–4291.

104 Wang, W., Yun, B.W., Kwon, E., Hong, J.K., Yoon, J., and Loake, G.J. (2006) *J. Exp. Bot.*, **57**, 1777–1784.

105 Grennan, A.K. (2007) *Plant Physiol.*, **144**, 1237–1239.

106 Kone, B.C., Kuncewicz, T., Zhang, W., and Yu, Z.Y. (2003) *Am. J. Physiol. Renal Physiol.*, **285**, 178–190.

6
Role of Nitric Oxide in Programmed Cell Death

Michela Zottini, Alex Costa, Roberto De Michele, and Fiorella Lo Schiavo

Summary

In recent years, nitric oxide has emerged as a key signaling molecule in plants. This small water- and lipid-soluble gas is involved in several biological processes including stomatal closure, seed germination, root development, and programmed cell death. Nitric oxide, together with reactive oxygen species and salicylic acid, plays a pivotal role in defense strategies against pathogen attack, such as the hypersensitive response and systemic acquired resistance. The cross talk among NO, ROS, and SA is not yet clear, with these molecules interacting at multiple levels, from their biosynthesis to signaling.
Because NO is involved in many types of PCD, such as that resulting from both biotic and abiotic stress, it has been proposed as a general stress molecule. In this chapter, the role of NO in cadmium-induced PCD will be discussed, suggesting a possible regulatory role in response to heavy metal stress.

6.1
Programmed Cell Death in Plants

Programmed cell death (PCD) is a genetically programmed process in which a cell guides its own destruction. PCD is responsible for removing redundant, misplaced, or damaged cells, thus contributing to both organism survival and development. Cell death provides developmental and biochemical flexibility to plants [1].

Almost all phases of plant life cycle, from germination through vegetative and reproductive development, are influenced by PCD. Several model systems such as barley aleurone protoplasts, *Zinnia elegans*, and *Arabidopsis* have been investigated.

Developmental uses of PCD include xylem differentiation [2, 3], deletion of embryonic suspensors [4], formation of functionally unisexual flowers from bisexual floral primordia [5, 6], root cap shedding [7], anther dehiscence [8], and leaf

Nitric Oxide in Plant Physiology. Edited by S. Hayat, M. Mori, J. Pichtel, and A. Ahmad

ISBN: 978-3-527-32519-1

morphogenesis [9]. Developmentally regulated PCD occurs at a predictable time and location and is induced by internal factors.

In contrast, environmentally induced PCD is initiated in response to external abiotic or biotic signals. One interesting example of PCD triggered by abiotic stress is aerenchyma formation in roots of plants subjected to hypoxia [10]. A cell death program is also activated in plants in response to cold, salt stress, and excessive UV irradiation. Another example of PCD in plants, as a response to stress, is triggered upon pathogen attack during the hypersensitive response (HR). This form of PCD is activated to surround the invading pathogen and inhibit its spread [11].

6.1.1
PCD Hallmarks and Regulation

A number of studies involving cell death in plants have observed striking similarities to the hallmark apoptotic features observed in animals, including DNA cleavage (ladders), DNA fragmentation (TUNEL positively reacting cells), involvement of caspase-like proteases, and formation of structures resembling apoptotic bodies [12]. In contrast, however, plant cells display several unique features compared to their animal counterparts, including a lack of "true" caspases [13], the presence of a rigid cell wall and, more importantly, the lack of an active phagocytosis system. Other unique features of plant cells include totipotency, chloroplasts, nonmotility, and numerous and sometimes large vacuoles harboring high levels of degradative enzymes.

In plants, PCD may take several forms, mostly reducible to two main types: apoptosis-like PCD and autophagic cell death [14]. The former type of death is a rapid process defined by rapid nuclear fragmentation, caspase-like activity, chromatin condensation followed by nuclear *blebbing* and DNA degradation. The latter type is a slow, highly regulated process that involves formation of the autophagosome, which sequesters part of the cytoplasm for nutrient recycling. During this process, new metabolic pathways are activated and others are turned off [15]. The HR shows not only some apoptotic features but also some autophagic ones, and, indeed, some are reminiscent of oncosis. It may, therefore, be more appropriate to consider HR a distinctive form of cell death, a view that will surely facilitate the integration of cytological aspects with other distinctive features of HR [11].

Genes that control PCD are conserved across wide evolutionary distances, defining a core set of biochemical reactions that are regulated by inputs from diverse upstream pathways [16]. These genes encode either antiapoptotic or proapoptotic proteins, which direct the cell to make life–death decisions. Recently, structural and functional homologues of the mammalian BAG (Bcl-2 athanogene) family were identified in *Arabidopsis* via bioinformatics to further illustrate the conservation of key PCD regulators [17]. Although displaying similarities, plant cells also exhibit distinctive features of PCD. The presence of chloroplasts, a prominent vacuole, and the cell wall are all unique aspects of plant cells and affect PCD responses accordingly [18]. Current theories speculate that chloroplasts, through the regulation of reactive oxygen species (ROS), may serve as a global messaging system in many plant PCD responses [19, 20]. Indeed, a range of abiotic and biotic stresses can raise ROS levels due to perturbations of chloroplastic and mitochondrial metabolism and defense responses to various pathogenic attacks, and failure to control ROS excess accumulation leads to oxidative stress and may further cell death [21].

Different types of molecules, such as endogenous plant growth regulators [22], fungal elicitors [23], reactive oxygen species [24], nitric oxide (NO) [25], salicylic acid (SA) [26], and others have been proposed to be involved in either induction or signal transduction of PCD. Among them, a very interesting role is played by NO.

6.2 NO as a Signaling Molecule

Nitric oxide is a small and highly reactive molecule that regulates a wide range of biological processes in phylogenetically distant species, and is involved in an impressive number of physiological processes as a signaling messenger. Many studies performed on animal systems have revealed the role of NO. Nitric oxide has also been reported to be involved in a variety of effects in arthropods, reptiles, amphibians, plants, fungi, and protists.

The past few years have seen an increasing number of studies dedicated to NO functions in plants. The emerging picture is that NO is a ubiquitous signaling molecule involved in diverse physiological processes including germination, root growth, stomatal closing, and adaptive responses to biotic and abiotic stresses [27]. Despite the growing evidence that NO is a key signaling molecule in plants, details of the mechanisms by which NO contributes to these processes are still lacking. Some studies have indicated the presence of NO target proteins and genes whose expression is regulated by NO, demonstrating that NO can directly influence the activity of plant proteins as well as the signaling cascade leading to gene expression [28].

6.2.1 NO Is Able to Induce or Inhibit PCD

Nitric oxide has now emerged, with oxidative stress, as a major arbiter of plant PCD; however, both cytotoxic and cytoprotecting/stimulating properties of NO have been described in plants. High levels of NO are associated with cell death and DNA fragmentation in *Taxus* cultures [29]. Exposure to NO has been demonstrated to reduce photosynthesis and to inhibit respiration in carrot cell cultures [30]. An increase in NO levels has also been associated with the progression of natural senescence and cytokinin-induced senescence [31], suggesting its involvement in the modulation of these physiological processes as well. On the other hand, exposure to low levels of NO improves the response of plants under diverse types of stresses. In addition, NO has been found to protect tomato plants from methylviologen damage by scavenging reactive oxygen species [32], confirming its antioxidant function. It has also been reported that NO plays a protective function in attenuating UV exposure damage. Nitric oxide has also been implicated in resistance to a virulent pathogen attack. After pathogen recognition, a complex signal transduction system triggers defense responses based on accumulation of ROS and NO.

6.2.2
Nitric Oxide and PCD in Hypersensitive Response

The role of NO in plants has been mainly studied, as a signal molecule involved in biotic stress, in the hypersensitive response triggered by pathogen attack. Nitric oxide was first detected in soybean cultures inoculated with virulent and nonvirulent bacteria. Furthermore, application of inhibitors of mammalian NOS suppressed HR cell death in *Arabidopsis thaliana* [25]. Many studies have provided evidence of a mammalian-type NO-responsive defense gene activation pathway in plants, which also involves cGMP-dependent components [33]. Subsequently, NO production has been demonstrated during the HR elicited in suspension cultures of *Arabidopsis* inoculated with *P. syringae* pv. *maculicola* [34] and tobacco cultures challenged with *P. syringae* pv. tomato [35]. During the HR, other signaling molecules besides NO are produced, such as reactive oxygen species and salicylic acid that are involved in plant resistance response. The relative concentrations of particular ROS and NO appear to be vital in the initiation or suppression of cell death. Delledonne *et al.* [36] showed that conditions favoring the accumulation of superoxide over H_2O_2 or an excess of NO reduces cell death. Application of NO donors has been shown to induce the expression of protective genes, such as alternative oxidase, which may aid in preventing the generation of ROS. ROS and NO are also involved in the regulation of SA biosynthesis [37].

Salicylic acid is implicated in the induction of programmed cell death associated with pathogen defense responses because SA levels increase in response to PCD-inducing infections, and PCD development can be inhibited by expression of salicylate hydroxylase encoded by the bacterial *nahG* gene [38]. Cell death can also be induced in *Arabidopsis* cell culture treated with SA. Treatment with 1 mM SA for 16 h killed 57% of *Arabidopsis* cells, while 0.5 mM SA did not affect cell viability. The cells treated with 1 mM SA showed characteristic cleavage of nuclear DNA into oligonucleosomal fragments, a hallmark of PCD (Figure 6.1).

The role of SA during pathogen attack is particularly relevant [39] both in promotion of a local response and in systemic acquired resistance (SAR) [40]. It is therefore important to decipher the SA signaling pathway to better understand the action of SA.

6.2.3
Signaling Component in SA-Induced NO Production

In a recent study, we demonstrated that SA activates NO synthesis in *Arabidopsis* and that this NO production proceeds, at least in part, through a NOS-dependent pathway [41]. NO production was detected by confocal microscopic analysis and spectrofluorometric assay in plant roots [41] and cultured cells (Figure 6.2). There was strong correlation between the two techniques and the two different biological

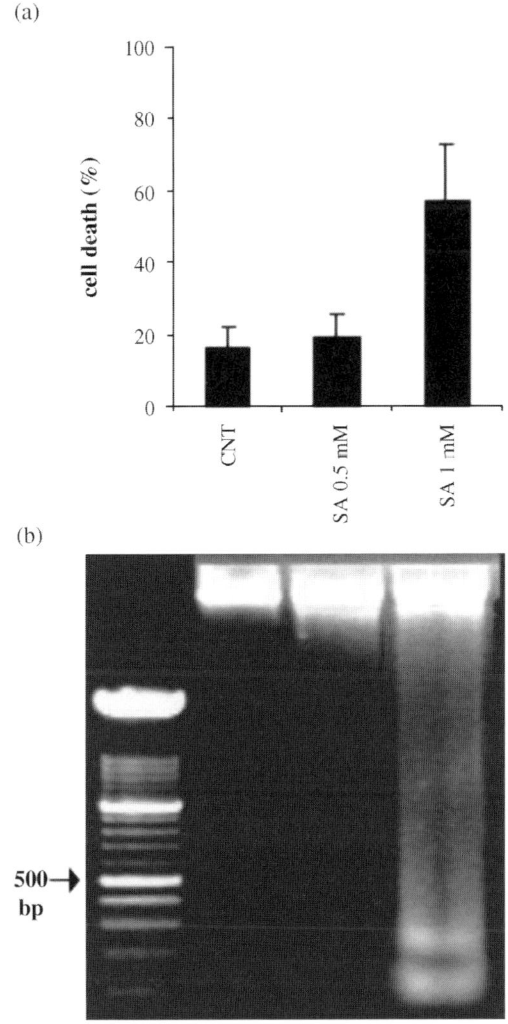

Figure 6.1 SA induces cell death in *Arabidopsis* suspension cultures. (a) Six-day-old cells were incubated with two different concentrations of SA (0.5 and 1 mM) and cell viability was estimated by Evan's blue staining at 16 h. Values represent mean + SE ($n = 6$). (b) Agarose gel analysis of DNA extracted from control and treated cells.

systems (i.e., cells and seedlings). This point strengthens the validity of the results, suggesting a broader occurrence of SA-induced NO production, not limited to specific tissues or organs. Moreover, the application of the EPR spectroscopy technology also confirmed these data. EPR spectroscopy combined with the spin-trap method has been identified as one technique for NO measurement, for determining the specificity of the EPR signal of nitrosylated adducts.

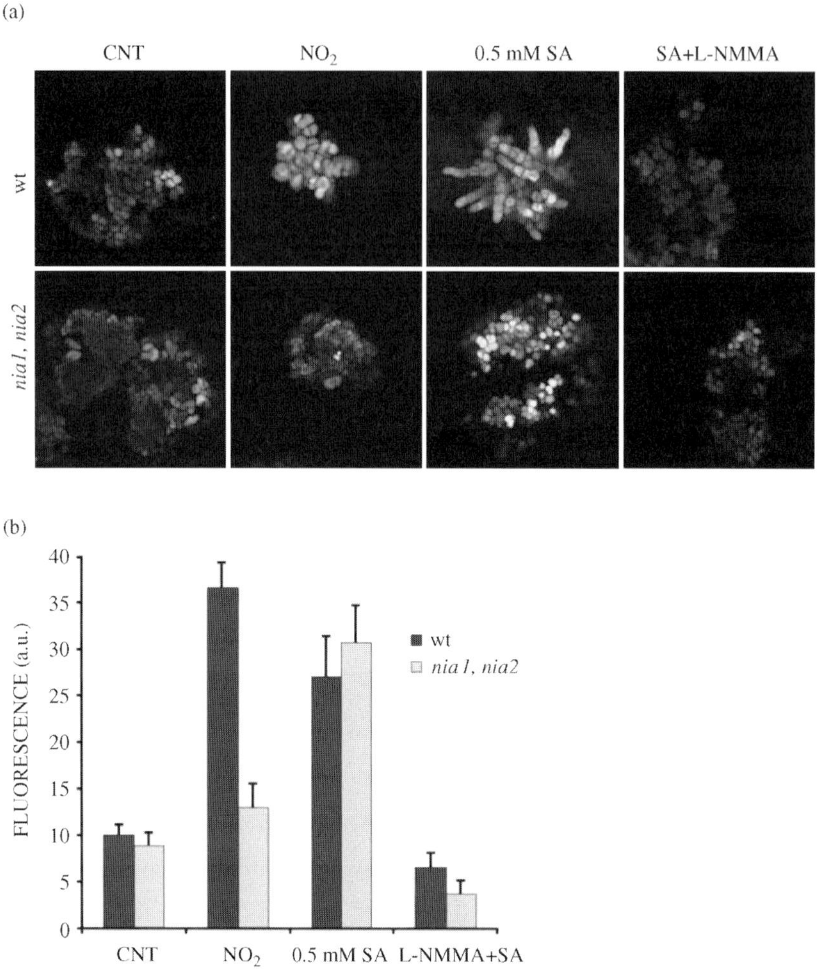

Figure 6.2 SA-induced NO production in *Arabidopsis* cell culture. This production is independent of nitrate reductase activity. In a nitrate reductase-deficient mutant, *nia1/nia2*, SA induces NO production that can be eliminated by pretreatment with the NOS inhibitor L-NMMA. (a) Fluorescence of DAF-FM DA loaded cells in response to 0.5 mM SA, 1 mM NO_2, or 0.5 mM SA + 1 mM L-NMMA. (b) Pixel intensity of DAF-FM DA fluorescence of cells.

Using protein kinase inhibitors, evidence has shown that phosphorylation events participate in the SA-induced signaling cascade leading to NO production. In particular, NO production completely depends on the activity of a specific casein kinase (CK2) since only a specific CK2 inhibitor (TBB [42]) was effective in both NO production and CK2 activity. This result is in accordance with previous data suggesting the involvement of CK2 in SA-induced activity [43]. In our work, we have also demonstrated that calcium is an early and crucial component of the SA signaling pathway. The role played by calcium in NO synthesis induced by SA may be that of a signal molecule and/or an enzymatic cofactor. Data from phosphorylation assay experiments allowed us to identify a few putative proteins whose phosphorylation is augmented by SA treatment and dependent on CK2 activity. Among them, a protein of 23 kDa has been identified (unpublished results) that shows similarity to an animal heat shock protein that is part of a chaperone system involved in several signaling pathways such as the regulation of NOS activity [44]. Our results clearly demonstrate that at least part of SA-induced NO synthesis occurs through a NOS-dependent route; calcium signaling and protein phosphorylation, through CK2, are early and essential components of the SA-induced pathway mediating NO synthesis. Our data also indicate the existence of a protein whose phosphorylation, increased by SA, depends on CK2 activity and the presence of calcium (Figure 6.3). These results, demonstrating the existence of an SA-dependent NO production in plants, are of interest because they suggest a regulatory loop capable of amplifying the signal involving NO and SA.

By better defining the relationship between SA and NO, these results contribute to a more detailed understanding of the metabolic pathways in which these molecules are involved. Finally, the SA-induced NO production model system could be a useful tool to identify the still unknown components of NOS-like activity. This is of great importance, considering that very recently it has been demonstrated that the activity of *NPR1*, a master regulator of SA-mediated defense genes, is regulated through *S*-nitrosylation [45]. This result is important and suggests a NO-dependent control mechanism of gene regulation in plant response to external stimuli. Further studies should be carried out to define the action mechanisms that link SA and PCD.

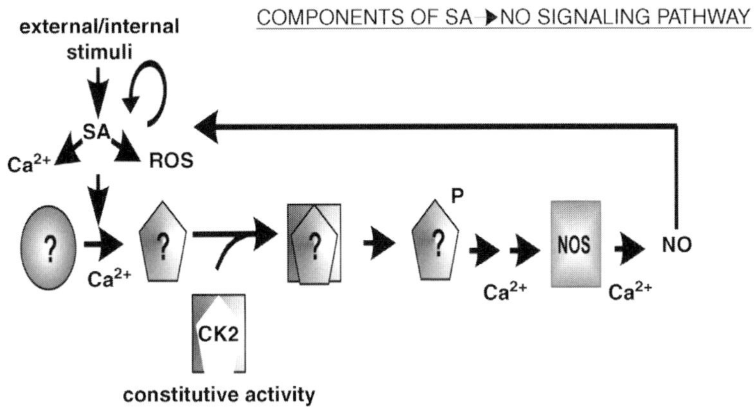

Figure 6.3 SA-induced NO synthesis occurs through a NOS-dependent route; calcium signaling and protein phosphorylation, through CK2, are early and essential components of the SA-induced pathway mediating NO synthesis.

6.3
Role of Mitochondria in NO-Induced PCD

Several data suggest that plant mitochondria play a central role in controlling programmed cell death pathways, representing both a stress sensor and dispatcher [46]. Cytochrome *c* and other factors released from mitochondria are considered to be important signals for triggering programmed cell death in animals as well as in plants [47].

It has been reported that NO imparts specific effects at the mitochondrial level, also *in vivo*, in carrot cell suspensions [30]; this is also true for *Arabidopsis* (unpublished results). NO decreases total respiration by about 50% in carrot cell suspensions treated with the NO donor sodium nitroprusside (SNP). This effect is associated with depolarization of the mitochondrial membrane potential and release of cytochrome *c* from mitochondria, suggesting a conserved signaling pathway in plants and animals. By using specific inhibitors for the cytochrome pathway (antimycin A) or the alternative pathway (SHAM), specific to these species, it has been demonstrated that NO specifically inhibits the former, while its activity is induced through the latter pathway. The induction of alternative oxidase activity is not merely due to a diversion of electrons, since cytochrome oxidase is inhibited, but is attributable to induction of protein expression [30]. This result has been confirmed by cDNA microarray analysis in *Arabidopsis*, where NO treatment induces the transcription of *Aox1a* within 3 h [48]. By preventing electron flux to cytochrome oxidase, prolonged exposure of mitochondria to NO causes the overreduction of the respiratory chain components and, consequently, an increase in O^{2-}, which augments $ONOO^-$ production and mitochondria damage [49]. The $ONOO^-$ induces in turn a mitochondrial Ca^{2+} efflux that promotes PTP formation, cytochrome *c* release, and thus caspase activation [50]. As yet, the link between cytochrome *c* release and the cell death mechanism is unclear because satisfactory plant gene homologues of Apaf1 are lacking and caspases have not been identified.

An intriguing example of abiotic stress in which the key role of mitochondria in PCD regulation can be envisaged is the signaling pathway activated by plant exposure to cadmium (Cd). Exposure to Cd leads to alterations in plant homeostasis and can result in cell death [51]. To date, little information is available about the signaling pathways involved in Cd-induced PCD. Several studies have focused on the importance of ROS production after treatment with high concentrations of Cd, which rapidly causes necrosis of exposed cells. Although Cd directly does not generate ROS by a Fenton reaction, there are indications that this metal inhibits antioxidant enzymes such as catalase, ascorbate peroxidase, superoxide dismutase, and glutathione reductase, inducing ROS production. In a recent paper [52], it has been demonstrated that the main ROS source in long-term Cd-treated plants are mitochondria, while plasma membrane NOX activity is responsible for Cd-induced H_2O_2 production during the initial phase of oxidative burst.

Upon Cd treatment, NO production was also detected in root tips of *Arabidopsis* [53]; a rapid NO burst has been observed within the first 6 h followed by a slower, gradual increase. The possibility that plant exposure to Cd might modulate NO production has been reported, but conflicting results on the impact of Cd on NO production have been published. Cadmium ions promote mROS formation by promoting overreduction of specific electron transport chain (ETC) components [52]. On the other hand, it is known that NO alters mitochondria functionality by inhibiting cytochrome oxidase. It would be interesting to investigate if the proapoptotic action of Cd is mediated by NO.

Cd-induced cell death can be interpreted as a model of PCD mediated by mitochondria in which both NO and ROS play crucial roles as signaling molecules.

6.4 Conclusions

We hypothesize that NO plays a key role in inducing cell death by interfering with mitochondria functionality, which can have a significant impact on the cellular balance between ROS generation and scavenging in both suspension cells and intact plants. This is of particular interest in leaves, given that another leaf organelle (the chloroplast) is often regarded as playing a dominant role in terms of both ROS generation and scavenging, suggesting an unexpectedly prominent role for mitochondria in these processes as well. On the other hand, it has recently been reported that during HR the ability of NO to induce cell death is due to its ability to act (along with cellular ROS) as a signaling molecule rather than targeting complex IV of the ETC [54].

The possibility of distinguishing among different "NO signatures," in terms of quantity, timing of production, and kinetics that differentially mark the responses to various stimuli, will be one of the main targets to pursue in future NO studies. Moreover, the comprehension of NO biosynthetic pathways induced in response to specific stimuli and the localization inside the cell where NO biosynthesis occurs, will be instrumental for understanding the different roles played by NO.

In order to gain insight into the action of NO in physiopathological responses in plants, it is essential to develop a reliable methodology for detecting and measuring NO with attention to quantification, compartmentalization, and kinetics of NO production.

References

1 Reape, T.J. and McCabe, P.F. (2008) *New Phytologist.*, **180**, 13–26.
2 Mittler, R. and Lam, E. (1995) *Plant Physiol.*, **108**, 489–493.
3 Fukuda, H. (2000) *Plant Mol. Biol.*, **44**, 245–253.
4 Giuliani, C., Consonni, G., Gavazzi, G., Colombo, M., and Dolfini, S. (2002) *Ann. Bot.*, **90**, 287–292.
5 Calderon-Urrea, A. and Della porta, S. (1999) *Development*, **126**, 435–441.
6 Caporali, E., Spada, A., Marziani, G., Failla, O., and Scienza, A. (2003) *Sex. Plant Reprod.*, **15**, 291–300.
7 Wang, H., Li, J., Bostock, R.M., and Gilchrist, D.G. (1996) *Plant Cell*, **8**, 375–391.
8 Bonner, L.J. and Dickinson, H.G. (1989) *New Phytol.*, **113**, 97–115.

9 Gunawardena, A. (2008) *J. Exp. Bot.*, **59**, 445–451.

10 Gunawardena, A., Pearce, D.M., Jackson, M.B., Hawes, C.R., and Evans, D.E. (2001) *Planta*, **212**, 205–214.

11 Mur, L.A.J., Kenton, P., Lloyd, A.J., Ougham, H., and Prats, E. (2008) *J. Exp. Bot.*, **59**, 501–520.

12 Li, W. and Dickman, M.B. (2004) *Biotechnol. Lett.*, **26**, 87–95.

13 Bonneau, L., Ge, Y., Drury, G.E., and Gallois, P. (2008) *J. Exp. Bot.*, **59**, 491–499.

14 Liu, Y., Schiff, M., Czymmek, K., Tallóczy, Z., Levine, B., and Dinesh-Kumar, S.P. (2005) *Cell*, **121**, 567–577.

15 Love, A.J., Milner, J.J., and Sadanandom, A. (2008) *Trends Plant Science*, **13**, 589–595.

16 Chae, H.J., Ke, N., Kim, H.R., Chen, S., Godzik, A., Dickman, M., and Reed, J.C. (2003) *Genes*, **323**, 101–113.

17 Doukhanina, E.V., Chen, S., van der Zalm, E., Godzik, A., Reed, J., and Dickman, M.B. (2006) *J. Biol. Chem.*, **281**, 18793–18801.

18 Hatsugai, N., Kuroyanagi, M., Nishimura, M., and Hara-Nishimura, I. (2006) *Apoptosis*, **11**, 905–911.

19 Samuilov, V.D., Lagunova, E.M., Dzyubinskaya, E.V., Izyumov, D.S., Kiselevsky, D.B., and Makarova, Y.V. (2002) *Biochemistry*, **67**, 627–634.

20 Zapata, J.M., Guera, A., Esteban-Carrasco, A., Martin, M., and Sabater, B. (2005) *Cell Death Differ*, **12**, 1277–1284.

21 Apel, K. and Hirt, H. (2004) *Annu. Rev. Plant. Biol.*, **55**, 373–399.

22 Hoeberichts, F.A. and Woltering, E.J. (2003) *Bioessays*, **25**, 47–57.

23 Hammond-Kosack, K. and Jones, J.D.G. (1996) *Plant Cell*, **8**, 1773–1791.

24 Levine, A., Tenhaken, R., Dixon, R., and Lamb, C. (1994) *Cell*, **79**, 583–593.

25 Delledonne, M., Xia, Y., Dixon, R.A., and Lamb, C. (1998) *Nature*, **394**, 585–588.

26 Dempsey, D.M.A., Shah, J., and Klessig, D.F. (1999) *Crit. Rev. Plant Sci.*, **18**, 547–575.

27 Besson-Bard, A., Pugin, A., and Wendehenne, D. (2008) *Annu. Rev. Plant. Biol.*, **59**, 21–39.

28 Lindermayr, C., Saalbach, G., and Durner, J. (2005) *Plant Physiol.*, **137**, 921–930.

29 Pedroso, M.C., Magalhaes, J.R., and Durzan, D. (2000) *J. Exp. Bot.*, **51**, 1027–1103.

30 Zottini, M., Formentin, E., Scattolin, M., Carimi, F., Lo Schiavo, F., and Terzi, M. (2002) *FEBS Lett.*, **515**, 75–77.

31 Carimi, F., Zottini, M., Costa, A., Cattelan, I., De Michele, R., Terzi, M., and Lo Schiavo, F. (2005) *Plant Cell Environ.*, **28**, 1171–1178.

32 Benigni, M.V. and Lamattina, L. (2001) *Plant Cell Environ.*, **24**, 267–728.

33 Durner, J., Wendehenne, D., and Klessig, D.F. (1998) *Proc. Natl. Acad. Sci. USA*, **95**, 10328–10333.

34 Clarke, A., Desikan, R., Hurst, R.D., Hancock, J.T., and Neill, S.J. (2000) *Plant J.*, **4**, 667–677.

35 Conrath, U., Amoroso, G., Kohle, H., and Sultemeyer, D.F. (2004) *Plant J.*, **38**, 1015–1922.

36 Delledonne, M., Zeier, J., Marocco, A., and Lamb, C. (2001) *Proc. Natl. Acad. Sci. USA*, **98**, 13454–13459.

37 Klessig, D.F., Durner, J., Noad, R., Navarre, D.A., Wendehenne, D., Kumar, D., Zhou, J.M., Shah, J., Zhang, S., Kachroo, P., Trifa, Y., Pontier, D., Lam, E., and Silva, H. (2000) *Proc. Natl. Acad. Sci. USA*, **97**, 8849–8855.

38 Brodersen, P., Malinovsky, F.G., Hematy, K., Newman, M.-A., and Mundy, J. (2005) *Plant Physiol.*, **138**, 1037–1045.

39 Yang, Y., Shah, J., and Klessig, D.F. (1997) *Gene. Dev.*, **11**, 1621–1639.

40 Alvarez, M.E. (2000) *PMB*, **44**, 429–444.

41 Zottini, M., Costa, A., De Michele, R., Ruzzane, M., Carimi, F., and Lo Schiavo, F. (2007) *J. Exp. Bot.*, **58**, 1397–1405.

42 Ruzzane, M., Penzo, D., and Pinna, L.A. (2002) *Biochem. J.*, **364**, 41–47.

43 Kang, H.G. and Klessig, D.F. (2005) *PMB*, **57**, 541–557.

44 Kadota, Y., Amigues, B., Ducassou, L., Madaoui, H., Ochsenbein, F., Guerois, R., and Shirasu, K. (2008) *EMBO Rep.* doi:

45 Tada, Y., Spoel, S.H., Pajerowska-Mukhtar, K., Mou, Z., Song, J., Wang, C.,

Zuo, J., and Dong, X. (2008) *Science*, **321**, 952–956.

46 Jones, A. (2000) *Trends Plant Sci.*, **5**, 225–230.

47 Yao, N., Eisfelder, B.J., Marvin, J., and Greenberg, J.T. (2004) *Plant J.*, **40**, 1000–1007.

48 Huang, X., von Rad, U., and Durner, J. (2002) *Planta*, **215**, 914–923.

49 Packer, M.A., Porteous, C.M., and Murphy, M.P. (1996) *Biochem. Mol. Biol. Int.*, **40**, 527–534.

50 Murphy, M.P. (1999) *Biochim. Biophys. Acta*, **1411**, 401–414.

51 Fojtova, M. and Kovařìk, A. (2000) *Plant Cell Environ.*, **22**, 531–537.

52 Heyno, E., Klose, C., and Krieger-Liszkay, A. (2008) *New Phytol.*, **179**, 687–699.

53 Bartha, B., Kolbert, Z., and Erdei, L. (2005) *Acta Biol. Szeg.*, **49**, 9–12.

54 Amirsadeghi, S., Robson, C.A., McDonald, A.E., and Vanlerberghe, G.C. (2006) *Plant Cell Physiol.*, **47**, 1509–1519.

7
Nitrate Reductase-Deficient Plants: A Model to Study Nitric Oxide Production and Signaling in Plant Defense Response to Pathogen Attack

Ione Salgado, Halley Caixeta de Oliveira, and Marcia Regina Braga

Summary

Nitric oxide has been recognized as a key signaling molecule in plant defense response against pathogens. Specific and general elicitors of plant defenses induce NO production in plant tissues and in cultured cells. Increased NO production activates the expression of defense-related genes, induces the hypersensitive response, and augments the production of antimicrobial compounds. The sources of NO during plant–pathogen interactions are still not fully understood. Several plant tissues produce NO through L-arginine oxidation by an activity similar to that of mammalian nitric oxide synthases. However, no gene or protein responsible for NO synthesis from L-arginine has been identified in plants. Nitrite has emerged as an alternative source of NO. In plants, the production of NO from nitrite may occur through various enzymatic and nonenzymatic mechanisms, some taking place in mitochondria. Nitrate reductase, a key enzyme for nitrate assimilation, may also reduce nitrite to NO as a secondary activity. Recent results from our laboratory suggest that NR is also important for L-arginine and nitrite production, the substrates for NO synthesis. This chapter focuses on how research on NR-deficient plants may contribute to the elucidation of mechanisms involved in NO production and signaling during plant–pathogen interactions.

7.1
Introduction

Plant resistance to microbial infection involves the activation of a wide variety of localized and systemic defenses that prevent pathogen replication and spread. Among these, the hypersensitive response (HR), characterized by rapid and localized

Nitric Oxide in Plant Physiology. Edited by S. Hayat, M. Mori, J. Pichtel, and A. Ahmad

ISBN: 978-3-527-32519-1

cell death, and the accumulation of antimicrobial compounds termed phytoalexins close to the infection site are believed to play important roles in plant resistance [1, 2]. Defensive responses can also be locally activated and subsequently expressed in parts of the plant far from the infection site, in a process known as systemic acquired resistance (SAR). Many pathogen-related (PR) proteins, which exhibit antimicrobial activity *in vitro* and *in vivo*, show increased synthesis in tissues distant from the site of microbial attack [3].

The induction of such defensive responses appears to be mediated by an initial recognition process between plants and pathogens, which involves the detection of microbial molecules, termed elicitors. However, a variety of chemically unrelated biotic and abiotic elicitors and environmental stresses can also activate resistance responses in plants, suggesting that the signaling pathways generated by different elicitors likely converge at one or several common events of transduction leading to the activation of common defensive strategies.

A substantial body of evidence indicates that H_2O_2 and salicylic acid (SA) are critical signaling molecules for local and systemic resistance [3]. More recently, data are emerging that demonstrate nitric oxide (NO) is another important signaling molecule in plant defenses against pathogens [4–8]. The production of NO, measured by diverse methods, increases during several plant–pathogen interactions [9–11], and different microbial compounds that induce defense responses in plants are also capable of inducing NO production [6, 12–17]. In addition, many genes involved in plant disease resistance are differentially expressed following exogenous application of NO [4, 18–20]. Moreover, the reduction in NO levels in plants carried out by different strategies impairs defensive responses [7, 8, 15].

Although these findings suggest a key role for NO in the signaling pathways of plant resistance against pathogen invasion, the sources of NO production in plants, including those stimulated during plant–pathogen interactions, are still not completely understood. Recently, the enzyme nitrate reductase (NR) has been implicated as an important component in plant response to pathogens and as providing NO for defensive purposes. The expression of NR genes and/or proteins is increased in potato tubers in response to *Phytophthora infestans* and to pathogen signals such as cell wall elicitor, arachidonic or salicylic acids [14, 21]. NR-deficient *Arabidopsis thaliana* plants have a decreased ability to synthesize NO and an increased susceptibility to *Pseudomonas syringae* [22, 23]. The silencing of NR genes in *Nicotiana benthamiana* also significantly decreases NO production induced by INF1, a fungal elicitin [16]. On the basis of these results, the researchers suggested that NR is involved in NO production for plant defense and that posttranslational modification of NR or the availability of nitrite may be rate-limiting steps in this process.

In addition to its important role in primary nitrogen metabolism [24], these findings suggest that NR may also be essential for plant resistance against pathogens. In this chapter, we provide an overview of the recent advances in the signaling role of NO in plant defense responses against microbial pathogens, the involvement of NR, and the use of NR-deficient plants for studying the mechanism of NO synthesis during plant–pathogen interactions.

7.2 Physicochemical Basis of NO Signaling

Besides its role in plant defense, NO has been implicated in recent years in several physiological processes in plants such as seed germination, root and leaf development, flowering, and stomatal movement, among others ([25, 26] and references therein). The physical and chemical properties of NO make it one of the most versatile signaling molecules, able to interact in different ways with distinct targets in the cell. NO is a gaseous free radical without charge and with a relatively long half-life (~5 s) compared to other radicals [27]. As one of the smallest diatomic molecules, it is highly diffusible and able to easily migrate through hydrophobic and hydrophilic compartments of the cell such as membranes and the cytosol [27]. Under physiological conditions, NO can be converted into other redox forms, being quickly oxidized to the nitrosonium cation (NO^+) or reduced forming the nitroxil anion (NO^-), which are important intermediaries in the biochemistry of NO [28].

NO is able to interact with transition metals such as ferrous iron within heme, leading to nitrosyl modifications of target molecules [29]. The reaction of NO with hemic Fe or with the Fe in Fe–S proteins is one important mechanism of NO action in the cell. In tobacco, aconitase, a Fe–S enzyme that catalyzes the isomerization of citrate into isocitrate, has its activity modified by NO [30]. In maize shoots, NO was shown to bind to the heme group of microsomal P450 and to interfere with the activity of cinnamic acid hydroxylase, a cytochrome P450-dependent monooxygenase, which is a key enzyme in the phenylpropanoid pathway [31]. In addition, it was recently suggested that NO activates NR through a direct interaction with the heme and molybdenum centers of the enzyme [32].

NO and NO^+ may cause *S*-nitrosylation of proteins by reacting with sulfur groups on cysteine residues [27]. At the cellular level, an important mechanism of NO-mediated signal transduction is the reversible *S*-nitrosylation of sulfhydryl groups on proteins that can alter the biological activity of these molecules. By using a proteomics approach, several polypeptides susceptible to *S*-nitrosylation have been identified in cultured *Arabidopsis* cells, including those related to the redox balance of the cell, and in metabolic, regulatory, and cytoskeleton proteins [33]. Recently, 16 *S*-nitrosylated proteins were identified in *Arabidopsis* plants during the progression of HR, suggesting a role of *S*-nitrosylation in plant resistance to avirulent pathogens [34]. *S*-Nitrosylation mediated by NO can cause diverse effects on the activity of proteins. For example, *S*-nitrosylation of type II peroxiredoxin, a crucial component of the antioxidant defense system, was reported to inhibit both peroxidase and peroxynitrite reductase activities of this enzyme [35]. Conversely, NO significantly activates NR by a mechanism that probably occurs via *S*-nitrosylation, in addition to its interaction with the heme and molybdenum centers of the enzyme [32].

NO can also react with thiol groups of low molecular weight compounds such as free cysteine and glutathione, forming *S*-nitrosoglutathione (GSNO) and nitrosocysteine (CysNO), respectively [36]. Although CysNO is less stable than GSNO, both *S*-nitrosothiols are much more stable than NO and can act as NO carriers [27, 33]. The observation that, in tobacco, GSNO induces the expression of *PAL* genes [5] and the systemic acquired resistance response against TMV [37] suggested that low molecular *S*-nitrosothiols could be important carriers of NO for defensive responses. The levels of GSNO, and consequently of NO, can be controlled *in vivo* by the *S*-nitrosoglutathione reductase (GSNOR) that is conserved from bacteria to humans [38]. In fact, manipulation of intracellular levels of GSNOR affects disease resistance in *A. thaliana*, highlighting the importance of this enzyme as a component of resistance signaling in plants [39]. Accordingly, *A. thaliana* transgenic plants with decreased amounts of GSNOR showed enhanced basal resistance against the biotrophic oomycete *Peronospora parasitica*, which was correlated with the higher levels of intracellular *S*-nitrosothiols and the constitutive activation of *PR-1*. Moreover, SAR was impaired in plants overexpressing GSNOR and enhanced in antisense plants, and it was correlated with changes in the *S*-nitrosothiol content both in local and in distant leaves [39].

Under aerobic conditions, NO can rapidly react with superoxide anion (O_2^-), producing peroxynitrite ($ONOO^-$) [40]. This mechanism of NO degradation was recently shown to play an important role in preventing the inhibitory effects of NO on plant mitochondrial respiration [41]. However, $ONOO^-$ excess can lead to the formation of nitrogen dioxide (NO_2) and hydroxyl radical (OH^-), a potent oxidant. Other reactions favored by peroxynitrite are the nitration of tyrosines (Tyr-NO_2) and the oxidation of thiol residues to sulfenic and sulfonic acids [25]. Increased protein nitration has been observed in tobacco leaves with antisense sequence to nitrite reductase (NiR) [42], in tobacco suspension cells treated with a fungal elicitor [43], and in olive leaves under salt stress [44], suggesting that $ONOO^-$ formation is stimulated under these conditions.

7.3
Defense Responses Mediated by NO

7.3.1
Accumulation of Defensive Compounds

Plants use a multifaceted array of defensive responses when confronted with potential pathogens. These responses are based on chemical and physical barriers designed to restrict pathogen entry and development within plant tissues. One of the best studied defense responses of plants to infection is the induced accumulation of low molecular weight antimicrobial compounds, termed phytoalexins. These substances are locally produced as a result of infection or stress, and there is strong evidence supporting their role in plant resistance to microbial attack [45]. In potato tubers, the early observation that NO supplied exogenously induced rishitin, a sesquiterpene phytoalexin [46], indicated that NO could mediate phytoalexin production. Soon, NO was shown to induce transcriptional activation of genes encoding phenylalanine ammonia-lyase (PAL) and chalcone synthase (CHS) [4], key enzymes in the phenylpropanoid pathway leading to the synthesis of phytoalexins in many plant species [47]. Indeed, NO is involved in the activation of the phenylpropanoid pathway as part of the soybean plant defense mechanism against attack by the fungus

Diaporthe phaseolorum f. sp. *meridionalis* (Dpm), the causal agent of stem canker disease [6]. The isoflavonoids daidzein and genistein were shown to accumulate in soybean cotyledons treated with Dpm extract and with sodium nitroprusside (SNP), an NO donor. The time course for the production of these metabolites was shorter when NO was provided directly, thus supporting the idea that the response of soybean cotyledons to Dpm elicitor involves endogenous NO formation, which triggers the biosynthesis of antimicrobial flavonoids [6]. Moreover, NO increased the activities of β-glucosidases, enzymes associated with deglucosylation of the precursors daidzin and genistin. This increase preceded the accumulation of these isoflavones in soybean cotyledons [48]. In the incompatible interaction between wheat and the stripe rust (*Puccinia striiformis*), the time course of NO production indicated that recognition of the pathogen was associated with two peaks of NO emission. The first one, detected during the early infection stage, was associated with increased PAL activity and plant resistance [49].

NO also mediates the elicitor-induced synthesis of saponins in ginseng cell culture [50]. The transcription of genes encoding squalene synthase and squalene epoxidase, two saponin synthesis enzymes, was induced by both oligogalacturonides and NO, but treatments that removed NO and inhibited NOS-like activity suppressed their transcription and saponin accumulation.

7.3.2 Hypersensitive Response

NO was proposed to play a fundamental role in the activation of the hypersensitive response in plants [4, 51]. The HR is a common host defense triggered at the site of microbial challenge and is characterized by a rapid and localized cell death, which prevents further pathogen spread [2, 52]. The generation of reactive oxygen species (ROS), including superoxide anion (O_2^-) and hydrogen peroxide (H_2O_2), is one of the earliest events in the HR, and this oxidative burst is thought to be a key component in coordinating plant responses to biotic and abiotic stress. In cultured soybean cells, NO acts synergistically with H_2O_2 to potentiate hypersensitive cell death [4, 51]. However, the synergism between ROS and NO is not a general rule among plant species since NO alone is sufficient to induce cell death in suspension cultures of *A. thaliana* when applied at concentrations similar to those induced by challenge with an avirulent bacterial strain [53].

The HR shares a subset of features with apoptotic cell death observed in animals (see Ref. [54]) and is, therefore, considered to be a form of programmed cell death. Different NO donors were shown to induce apoptotic-like cell death in citrus [55], carrot [56], and *Arabidopsis* [53] cells and in leaves and calluses of *Kalanchoë daigremontiana* and *Taxus brevifolia* [57, 58]. These findings point to NO as the signal responsible for triggering programmed cell death during the HR in plant–pathogen interactions. Kinetic analysis of NO production and cell death after inoculation of *A. thaliana* with an avirulent bacteria showed that NO accumulation paralleled HR progression, suggesting that NO could act as an intercellular signal essential for the development of the HR rather than controlling HR triggering [59].

In oat plants, it was proposed that NO and ROS play a role in HR progression inducing apoptosis in adjacent cells, since treatments that remove these radicals suppressed the morphological features of apoptotic cell death induced by an avirulent fungus, such as heterochromatin condensation and DNA laddering [60]. Recent results reported by Ali *et al.* [61] suggest that activation of cyclic nucleotide gated channel inward Ca^{2+} current is linked to NO generation and HR in *Arabidopsis*, by increases in cytosolic levels of Ca^{2+}/calmodulin that activates a NOS-like enzyme.

7.3.3
Systemic Responses

Endogenous salicylic acid is considered to be a critical signaling molecule in the pathways leading to SAR, since it induces PR genes [62]. The observation that NO induces the expression of defense-related genes for PR proteins and the synthesis of SA in tobacco cells [5] suggested a role for this radical in SAR. SA also enhances NO production in soybean leaves [63], indicating that both signaling molecules can be self-amplified and may act synergistically to activate local and systemic defense responses as observed in tobacco plants [37].

Plant wounding caused by pathogens or insect attack can lead to the systemic production of proteinase inhibitor proteins (PIPs), which limit protein digestion and retard both growth and development of pests and pathogens [64]. In tomato leaves, the synthesis of PIP I is inhibited by NO donors. NO also blocks H_2O_2 production and PIP synthesis, promoted by systemin, oligogalacturonides, and jasmonic acid, three powerful inducers of the systemic accumulation of these proteins in Solanaceae species [65]. NO appears to inhibit the expression of the wound-inducible defense gene by downregulating the pathway leading to jasmonic acid synthesis. Jasmonic acid has been directly implicated in the signal transduction pathway leading to the induction of defensive genes in plants in response to insect and pathogen attacks [66].

7.3.4
Stomatal Closure

Recently, Melotto *et al.* [67] provided evidence that stomatal closure in *A. thaliana* is part of the plant innate resistance response to restrict bacterial invasion. Unlike fungal pathogens, bacteria lack the ability to directly penetrate the plant epidermis, and, therefore, stomata serve as natural openings for their entry. Stomatal closure, a physiological response induced by abcisic acid (ABA), is effected through a complex intracellular signaling cascade that involves SA and NO. Removal of NO by scavengers inhibits the stomatal closure induced by ABA, and stomatal closure is reduced in NO-deficient mutants (see Ref. [68]). Melotto *et al.* [67] showed that although avirulent and virulent bacterial strains have the ability to cause NO-mediated stomatal closure, only the virulent strain causes stomatal reopening as a pathogenesis strategy. These findings suggest that NO production in guard cells induced by bacterial elicitors may be crucial to the resistance of *A. thaliana* to *P. syringae* infection.

7.4
Substrates for NO Production During Plant–Pathogen Interactions

7.4.1
Production of NO from L-Arginine

In animals, it is well established that a family of enzymes called nitric oxide synthases (NOSs) is the primary system responsible for NO production. The NOSs catalyze the formation of NO and L-citrulline by a five-electron oxidation of the guanidine nitrogen of the amino acid L-arginine in a reaction dependent on O_2, NADPH, heme, tetrahydrobiopterin, calmodulin, FAD, and FMN [69]. In mammals, the macrophagic NOS, the main isoform involved in defense against infection, is induced during the immune response [70].

In plants, NO production sensitive to NOS inhibitors was first detected in soybean cell suspensions infected with *P. syringae* [4]. In addition, NOS activity, measured by the conversion of radiolabeled L-arginine to L-citrulline, was shown to increase in tobacco leaves infected with the tobacco mosaic virus [5] and in soybean cotyledons after treatment with an elicitor from the fungus *D. phaseolorum* [6]. Although an NOS-like activity has been observed in many plant tissues (see Ref. [71]), genome sequencing of *A. thaliana* [72] has not revealed any gene or protein sequence similar to animal NOS. It was proposed that plants could produce NO through an enzyme different from the mammalian enzyme. In 2003, Guo *et al.* [73] identified a gene in *A. thaliana* that encodes a protein homologous to the proposed enzyme for NO production in nervous tissue of the snail *Helix pomatia* [74]. This gene was initially called *AtNOS1* because it would codify a protein (AtNOS1) that converts L-arginine into L-citrulline [73]. However, Zemojtel *et al.* [75] were unable to detect NOS activity in the recombinant protein encoded by the *AtNOS1* gene. Crawford's group subsequently renamed this gene *NOA1 – NO-associated 1* [76]. The *Atnoa1* mutant produces reduced amounts of NO compared to wild-type genotype [73] and develops more severe disease symptoms after infection with *P. syringae* pv. tomato [15], suggesting that NOA1-dependent NO synthesis is involved in defense against this pathogen.

7.4.2
Production of NO from Nitrite

Nitrite has emerged as an alternative source of NO in plants. The production of NO from nitrite may occur through various enzymatic and nonenzymatic mechanisms. In acidic environments, the chemical reduction of nitrite to HNO_2 is favored, and NO is formed through a direct electron transfer reaction that occurs in the presence of reductants such as ascorbic acid and phenols (see Ref. [77]). The nonenzymatic reduction of nitrite in apoplastic acidic conditions was demonstrated in the aleurone layer of barley [78].

Electrons flowing from the respiratory chain in mammalian mitochondria were shown to reduce nitrite to NO [79]. This mitochondrial nitrite-reducing activity was also demonstrated in tobacco leaves, tobacco cell suspensions [80], and *A. thaliana* leaf homogenates [22]. More recently, the ability of chloroplasts to synthesize NO from nitrite was described in soybean leaves, but the mechanism responsible for this synthesis has not been elucidated [81].

NO synthesis by a nitrite-NO oxidoreductase (Ni-NOR) activity, found exclusively in root plasma membrane of *N. tabacum*, has been described [82]. The involvement of Ni-NOR in physiological root processes, including development, response to anoxia, and symbiosis, has been suggested [83]; however, the genetic identity of Ni-NOR is not known.

Nitrate reductase, a key enzyme for nitrate assimilation in plants that reduces nitrate to nitrite by an NAD(P)H-dependent mechanism, may also reduce nitrite to NO as a secondary activity. Nitrite is normally reduced to ammonium by the nitrite reductase enzyme. However, under some circumstances, NR can additionally catalyze the monoelectronic reduction of nitrite resulting in the production of NO [84, 85]. In *Arabidopsis* leaf extracts, NR-dependent NO production appears to be as high as or even much higher than NOS-dependent NO emission [86].

The nitrite-reducing activity of NR has been supported by the observation of lowered NO emission in NR-deficient organisms. The cc-2929 NR mutant of the green unicellular alga *Chlamydomonas reinhardtii* displays a much lower NO production than wild-type cells [87]. *N. benthamiana* protoplasts with silenced NR genes produce less NO in response to elicitin treatment [16]. Moreover, in *A. thaliana*, NO production induced by *Verticillium dahliae* toxin is much higher in wild-type plants in comparison to the *nia1 nia2* mutant, which is deficient in the two structural genes for NR [88]. Many hormonal stimuli have also been shown to induce an NR-associated NO synthesis (see Ref. [25]). For instance, roots of the *nia1 nia2* mutant do not produce NO after treatment with auxin, a response clearly seen in wild-type plants [89]. In addition, in this mutant, ABA neither stimulates NO production by guard cells nor induces stomatal closure [90].

Although the *nia1 nia2* mutant has lower nitrite and NO in comparison to wild-type plants, it is still able to produce this radical when nitrite is exogenously supplied [22]. It has been proposed that the reduced NO production from nitrite in NR-deficient plants results from their low endogenous contents of nitrite [22]. Accordingly, plants or cell suspension cultures of tobacco-free NR exposed for short periods with nitrate do not emit NO. However, when nitrite is provided, these cells are able to emit NO with the same velocity as cells expressing NR [80], indicating that nitrite more than nitrate is the substrate required for NO production. Indeed, several studies have indicated that NO production by plants is linked to the endogenous level of nitrite. In tobacco leaves with NiR antisense, and, therefore, with very low NiR activity, the concentration of nitrite is 10 times higher, and NO emission is 100 times higher, compared to wild type [42].

NR is tightly regulated at the transcriptional and posttranscriptional levels. A rapid and reversible phosphorylation of a conserved serine residue modifies NR activity in response to carbon and light signals. In the presence of cations and polyamines, the phosphorylated form of NR interacts with 14-3-3 proteins, resulting in NR inactivation. Rapid downregulation of NR avoids accumulation of toxic nitrite [91, 92]. In *N. plumbaginifolia*, the expression of a modified tobacco NR in which the regulatory serine has been changed into aspartic acid results in a permanently active NR enzyme [92]. Leaf tissues and root segments of this transgenic *N. plumbaginifolia* with constitutively high NR activity show increased nitrite formation and NO emission [93].

7.5 The Role of Nitrate Reductase in NO Production During Plant–Pathogen Interactions

Recent results indicate that the activity of NR is important for proper plant defense responses against pathogen attack. The mutant *nia1 nia2* of *A. thaliana* possesses a reduced capacity for NO synthesis in response to inoculation with an avirulent strain of *P. syringae* pv. *maculicola* [22]. However, when nitrite was infiltrated in *nia1 nia2* leaves, the production of NO in response to pathogen inoculation was rescued, indicating that the inability of NR-deficient plants to produce NO is due to their low endogenous levels of nitrite [22]. Furthermore, analysis of cell death in *A. thaliana* during its interaction with *P. syringae* showed that NR-deficient plants lost their ability to develop a normal HR at the site of bacterial inoculation, a response clearly seen in plants expressing a functional NR [23]. Since NO is essential for the induction of defense-associated genes, the inability of *nia1 nia2* plants to develop a normal HR to *P. syringae* was suggested to result from their failure to promptly produce this radical [23].

The positive correlation between NO content and resistance to pathogens has also been demonstrated by using other approaches. For instance, introduction of the NO dioxygenase gene, which is related to NO degradation, in *A. thaliana* and in an avirulent strain of *P. syringae* resulted in reduced NO levels *in planta*, preventing HR induction and expression of defense genes [7]. A similar approach was used by Boccara *et al.* [8] who introduced the flavohemoglobin HmpX gene in *P. syringae* and observed a negative effect on hypersensitive cell death induction in soybean cell suspensions. On the other hand, the HmpX-deficient mutant *Erwinia chrysanthemi* led to an unusually high accumulation of NO in *Saintpaulia ionantha* plants followed by the HR. These responses were not observed when a strain expressing the NO-degradation protein was the inoculum [8].

The *nia1 nia2* mutant of *A. thaliana* has a low content of amino acids in its leaves [23] as a consequence of its limited capacity for nitrogen assimilation. L-Arginine is about 10 times lower in the *nia1 nia2* mutant than in the wild type [23]. These findings suggest that the reduced NO-producing capacity of NR-deficient plants could result not only from the lower content of nitrite but also from the decreased levels of L-arginine, highlighting the important role NR plays in providing the substrates for NO generation in plants (see Ref. [94]).

Compromised nitrogen assimilation in NR-deficient *A. thaliana* could affect physiological processes, such as plant response to pathogens, irrespective of low NO production. However, when *nia1 nia2* mutants were irrigated with L-glutamine or L-arginine, their total amino acid content was recovered, and the wild phenotype was rescued. Despite this, amino acid-recovered *nia1 nia2* plants were unable to increase NO emission and to develop an HR to *P. syringae*, showing severe disease symptoms with intense bacterial growth in leaves [95]. These results indicated that the impaired resistance response of *nia1 nia2 A. thaliana* to *P. syringae* seems not to be related to the low endogenous amino acid content and also that NO produced from nitrite, instead of L-arginine, is necessary for a proper response to pathogen attack. This is in agreement with the fact that when nitrite is exogenously infiltrated into *A. thaliana* leaves, HR to *P. syringae* is recovered [23].

In conclusion, although substantial evidence indicates that L-arginine is important for NO production in plants, our results with an NR-deficient mutant of *A. thaliana* suggest that nitrite seems to be the crucial source of NO in plant–pathogen interactions. However, the importance of nitrite for NO generation during the plant response to microbial attack requires further investigation, in particular the examination of altered gene expression and changes in protein activity. Although it has recently been shown that nitrite can alter gene expression independent of NO [96], it is important to consider that increased nitrite-reducing activity during plant–pathogen interactions can lead to altered expression of NO-responsive genes. Moreover, enhanced NO production can cause *S*-nitrosylation of proteins, including NR [32], resulting in increased nitrite levels, the relevant substrate for NO production during plant–pathogen interactions.

References

1 Paxton, J.D. (1991) Biosynthesis and accumulation of legume phytoalexins, in *Mycotoxinas and Phytoalexins* (eds R.P. Sharma and D.K. Shalunke), CRC Press, Boca Raton, FL, pp. 485–499.

2 Lamb, C. and Dixon, R.A. (1997) *Annu. Rev. Plant Physiol. Plant Mol. Biol.*, **48**, 251–275.

3 Klessig, D.F., Durner, J., Noad, R., Navarre, D.A., Wendehenne, D., Kumar, D., Zhou, J.M., Shah, J., Zhang, S., Kachroo, P., Trifa, Y., Pontier, D., Lam, E., and Silva, H. (2000) *Proc. Natl. Acad. Sci. USA*, **97**, 8849–8855.

4 Delledonne, M., Xia, Y., Dixon, R.A., and Lamb, C. (1998) *Nature*, **394**, 585–588.

5 Durner, J., Wendehenne, D., and Klessig, F. (1998) *Proc. Natl. Acad. Sci. USA*, **9**, 10328–10333.

6 Modolo, L.V., Cunha, F.Q., Braga, M.R., and Salgado, I. (2002) *Plant Physiol.*, **130**, 1288–1297.

7 Zeier, J., Delledonne, M., Mishina, T., Severi, E., Sonoda, M., and Lamb, C. (2004) *Plant Physiol.*, **136**, 2875–2886.

8 Boccara, M., Mills, C.E., Zeier, J., Anzi, C., Lamb, C., Poole, R.K., and Delledonne, M. (2005) *Plant J.*, **43**, 226–237.

9 Romero-Puertas, M.C., Perazzolli, M., Zago, E.D., and Delledonne, M. (2004) *Cell Microbiol.*, **6**, 795–803.

10 Salgado, I., Saviani, E.E., Modolo, L.V., and Braga, M.R. (2004) Nitric oxide signaling in plant defence responses to pathogen attack, in *Advances in Plant Physiology*, vol. 7 (ed. A. Hemantaranjan), Scientific Publishers, Jodhpur, India, pp. 117–137.

11 Vandelle, E. and Delledonne, M. (2008) *Methods Enzymol.*, **437**, 575–594.

12 Foissner, I., Wendehenne, D., Langebartels, C., and Durner, J. (2000) *Plant J.*, **23**, 817–824.

13 Lamotte, O., Gould, K., Lecourieux, D., Sequeira-Legrand, A., Lebun-Garcia, A.,

Durner, J., Pugin, A., and Wendehenne, D. (2004) *Plant Physiol.*, **135**, 516–529.
14 Yamamoto, A., Katou, S., Yoshioka, H., Doke, N., and Kawakita, K. (2004) *J. Gen. Plant Pathol.*, **70**, 85–92.
15 Zeidler, D., Zahringer, U., Gerber, I., Dubery, I., Hartung, T., Bors, W., Hutzler, P., and Durner, J. (2004) *Proc. Natl. Acad. Sci. USA*, **101**, 15811–15816.
16 Yamamoto-Katou, A., Katou, S., Yoshioka, H., Doke, N., and Kawakita, K. (2006) *Plant Cell Physiol.*, **47**, 726–735.
17 Besson-Bard, A., Griveau, S., Bedioui, F., and Wendehenne, D. (2008) *J. Exp. Bot.* doi: 10. 1093/jxb/ern189.
18 Polverari, A., Molesini, B., Pezzotti, M., Buonaurio, R., Marte, M., and Delledonne, M. (2003) *Mol. Plant–Microbe Interact.*, **16**, 1094–1105.
19 Ferrarini, A., De Stefano, M., Baudouin, E., Pucciariello, C., Polverari, A., Puppo, A., and Delledonne, M. (2008) *Mol. Plant–Microbe Interact.*, **21**, 781–790.
20 Palmieri, M.C., Sell, S., Huang, X., Scherf, M., Werner, T., Durner, J., and Lindermayr, C. (2008) *J. Exp. Bot.*, **59**, 177–186.
21 Yamamoto, A., Katou, S., Yoshioka, H., Doke, N., and Kawakita, K. (2003) *J. Gen. Plant Pathol.*, **69**, 218–229.
22 Modolo, L.V., Augusto, O., Almeida, I.M., Magalhaes, J.R., and Salgado, I. (2005) *FEBS Lett.*, **579**, 3814–3820.
23 Modolo, L.V., Augusto, O., Almeida, I.M.G., Pinto-Maglio, C.A.F., Oliveira, H.C., Seligman, K., and Salgado, I. (2006) *Plant Sci.*, **171**, 34–40.
24 Crawford, N.M., Kahn, M.L., Leustek, T., and Long, S.R. (2000) Nitrogen and sulfur, in *Biochemistry and Molecular Biology of Plants* (eds B.B. Buchanan, W. Gruissem, and R.L. Jones), American Society of Plant Physiologists, Rockville, MD, pp. 786–849.
25 Lamattina, L., Garcia-Mata, C., Graziano, M., and Pagnussat, G. (2003) *Annu. Rev. Plant Biol.*, **54**, 109–136.
26 Wilson, I.D., Neill, S.J., and Hancock, J.T. (2008) *Plant Cell Environ.*, **31**, 622–631.
27 Stamler, J.S., Singel, D.J., and Loscalzo, J. (1992) *Science*, **258**, 1898–1902.
28 Hughes, M.N. (1999) *Biochim. Biophys. Acta*, **1411**, 263–272.
29 Stamler, J.S., Lamas, S., and Fang, F.C. (2001) *Cell*, **106**, 675–683.
30 Navarre, D., Wendenhenne, D., Durner, J., Noad, R., and Klessing, D.F. (2000) *Plant Physiol.*, **122**, 573–582.
31 Enkhardt, U. and Pommer, U. (2000) *J. Appl. Bot.*, **74**, 151–154.
32 Du, S., Zhang, Y., Lin, X., Wang, Y., and Tang, C. (2008) *Plant Cell Environ.*, **31**, 195–204.
33 Lindermayr, C., Saalbach, G., and Durner, J. (2005) *Plant Physiol.*, **137**, 921–930.
34 Romero-Puertas, M.C., Campostrini, N., Mattè, A., Righetti, P.G., Perazzolli, M., Zolla, L., Roepstorff, P., and Delledonne, M. (2008) *Proteomics*, **8**, 1459–1469.
35 Romero-Puertas, M.C., Laxa, M., Mattè, A., Zaninotto, F., Finkemeier, I., Jones, A.M., Perazzolli, M., Vandelle, E., Dietz, K.J., and Delledonne, M. (2007) *Plant Cell*, **19**, 4120–4130.
36 Vanin, A.F. (1995) *Biochemistry (Moscow)*, **60**, 441–447.
37 Song, F. and Goodman, R.M. (2001) *Mol. Plant–Microbe Interact.*, **14**, 1458–1462.
38 Liu, L., Hausladen, A., Zeng, M., Que, L., Heitman, J., and Stamler, J.S. (2001) *Nature*, **410**, 490–494.
39 Rustérucci, C., Espunya, M.C., Díaz, M., Chabannes, M., and Martínez, M.C. (2007) *Plant Physiol.*, **143**, 1282–1292.
40 Radi, R., Cassina, A., Hodara, R., Quijano, C., and Castro, L. (2002) *Free Radic. Biol. Med.*, **33**, 1451–1464.
41 de Oliveira, H.C., Wulff, A., Saviani, E.E., and Salgado, I. (2008) *Biochim. Biophys. Acta*, **1777**, 470–476.
42 Morot-Gaudry-Talarmain, Y., Rockel, P., Moureaux, T., Quillere, I., Leydecker, M.T., Kaiser, W.M., and Morot-Gaudry, J.F. (2002) *Planta*, **215**, 708–715.
43 Saito, S., Yamamoto-Katou, A., Yoshioka, H., Doke, N., and Kawakita, K. (2006) *Plant Cell Physiol.*, **47**, 689–697.

44 Valderrama, R., Corpas, F.J., Carreras, A., Fernández-Ocaña, A., Chaki, M., Luque, F., Gómez-Rodríguez, M.V., Colmenero-Varea, P., del Río, L.A., and Barroso, J.B. (2007) *FEBS Lett.*, **58**, 453–461.

45 Hammerschmidt, R. (1999) *Annu. Rev. Phytopathol.*, **37**, 285–306.

46 Noritake, T., Kawakita, K., and Doke, N. (1996) *Plant Cell Physiol.*, **37**, 113–116.

47 Smith, C.J. (1996) *New Phytol.*, **132**, 1–45.

48 Kretzschmar, F.S., Aidar, M.P., Salgado, I., and Braga, M.R. (2009) *Environ. Exp. Bot.*, **65**, 319–329.

49 Guo, P., Cao, Y., Li, Z., and Zhao, B. (2004) *Plant Cell Environ.*, **27**, 473–477.

50 Hu, X., Neill, S.J., Cai, W., and Tang, Z. (2003) *Funct. Plant Biol.*, **30**, 901–907.

51 Delledonne, M., Zeier, J., Marocco, A., and Lamb, C. (2001) *Proc. Natl. Acad. Sci. USA*, **98**, 13454–13459.

52 Dangl, J.L., Dietrich, R.A., and Richberg, M.H. (1996) *Plant Cell*, **8**, 1793–1807.

53 Clarke, A., Desikan, R., Hurst, R.D., Hancock, J.T., and Neill, S.J. (2000) *Plant J.*, **24**, 667–677.

54 Gilchrist, D.G. (1998) *Annu. Rev. Phytopathol.*, **36**, 393–414.

55 Saviani, E.E., Orsi, C.H., Oliveira, J.F., Pinto-Maglio, C.A., and Salgado, I. (2002) *FEBS Lett.*, **510**, 136–140.

56 Zottini, M., Formentin, E., Scattolin, M., Carimi, F., Lo Schiavo, F., and Terzi, M. (2002) *FEBS Lett.*, **515**, 75–78.

57 Pedroso, M.C., Magalhaes, J.R., and Durzan, D. (2000) *Plant Sci.*, **157**, 173–180.

58 Pedroso, M.C., Magalhaes, J.R., and Durzan, D. (2000) *J. Exp. Bot.*, **51**, 1027–1036.

59 Zhang, C., Czymmek, K.J., and Shapiro, A.D. (2003) *Mol. Plant–Microbe Interact.*, **16**, 962–972.

60 Tada, Y., Mori, T., Shinogi, T., Yao, N., Takahashi, S., Betsuyaku, S., Sakamoto, M., Park, P., Nakayashiki, H., Tosa, Y., and Mayama, S. (2004) *Mol. Plant–Microbe Interact.*, **17**, 245–253.

61 Ali, R., Ma, W., Lemtiri-Chlieh, F., Tsaltas, D., Leng, Q., von Bodman, S., and Berkowitz, G.A. (2007) *Plant Cell*, **19**, 1081–1095.

62 Malamy, J. and Klessig, D.F. (1992) *Plant J.*, **2**, 643–654.

63 Klepper, L.A. (1991) *Pest Biochem. Physiol.*, **39**, 43–48.

64 Ryan, C.A. (1990) *Annu. Rev. Phytopathol.*, **28**, 425–449.

65 Orozco-Cárdenas, M.L. and Ryan, C.A. (2002) *Plant Physiol.*, **130**, 487–493.

66 Heil, M. and Bostock, R.M. (2002) *Ann. Bot.*, **89**, 503–512.

67 Melotto, M., Underwood, W., Koczan, J., Nomura, K., and He, S.Y. (2006) *Cell*, **126**, 969–980.

68 Neill, S., Barros, R., Bright, J., Desikan, R., Hancock, J., Harrison, J., Morris, P., Ribeiro, D., and Wilson, I. (2008) *J. Exp. Bot.*, **59**, 165–176.

69 Stuehr, D.J. (1997) *Annu. Rev. Pharmacol. Toxicol.*, **37**, 339–359.

70 MacMicking, J., Xie, Q.W., and Nathan, C. (1997) *Annu. Rev. Immunol.*, **15**, 323–350.

71 del Rio, L.A., Corpas, F.J., and Barroso, J.B. (2004) *Phytochemistry*, **65**, 783–792.

72 The Arabidopsis Genome Initiative (2000) *Nature*, **408**, 796–815.

73 Guo, F., Okamoto, M., and Crawford, N.M. (2003) *Science*, **302**, 100–103.

74 Huang, S., Kerschbaum, H.H., Engel, E., and Hermann, A. (1997) *J. Neurochem.*, **69**, 2516–2528.

75 Zemojtel, T., Froblich, A., Palmicri, M.C., Kolanczyk, M., Mikula, I., and Wyrwicz, L.S. (2006) *Trends Plant Sci.*, **11**, 524–525.

76 Crawford, N.M., Galli, M., Tischner, R., Heimer, Y.M., Okamobo, M., and Mack, A. (2006) *Trends Plant Sci.*, **11**, 526–527.

77 Yamasaki, H., Sakihama, Y., and Takahashi, S. (1999) *Trends Plant Sci.*, **4**, 128–129.

78 Bethke, P.C., Badger, M.R., and Jones, R.L. (2004) *Plant Cell*, **16**, 332–341.

79 Kozlov, A.V., Staniek, K., and Nohl, H. (1999) *FEBS Lett.*, **454**, 127–130.

80 Planchet, F., Kapuganti, J.G., Sonoda, M., and Kaiser, W.M. (2005) *Plant J.*, **41**, 732–743.

81 Jasid, S., Simontacchi, M., Bartoli, C.G., and Puntarulo, S. (2006) *Plant Physiol.*, **142**, 1246–1255.

82 Stohr, C., Strule, F., Marx, G., Ullrich, W.R., and Rockel, P. (2001) *Planta*, **212**, 835–841.

83 Stohr, C. and Stremlau, S. (2006) *J. Exp. Bot.*, **57**, 463–470.

84 Yamasaki, H. (2000) *Philos. Trans. R. Soc. Lond. B Biol. Sci.*, **355**, 1477–1488.

85 Rockel, P., Strube, F., Rockel, A., Wildt, J., and Kaiser, W.M. (2002) *J. Exp. Bot.*, **53**, 1–8.

86 Meyer, C., Lea, U.S., Provan, F., Kaiser, W.M., and Lillo, C. (2005) *Photosynth. Res.*, **83**, 181–189.

87 Sakihama, Y., Nakamura, S., and Yamasaki, H. (2002) *Plant Cell Physiol.*, **43**, 290–297.

88 Shi, F.M. and Li, Y.Z. (2008) *BMB Rep.*, **41**, 79–85.

89 Kolbert, Z., Bartha, B., and Erdei, L. (2008) *J. Plant Physiol.*, **165**, 967–975.

90 Desikan, R., Griffiths, R., Hancock, J., and Neill, S. (2002) *Proc. Natl. Acad. Sci. USA*, **99**, 16314–16318.

91 Riens, B. and Heldt, H.W. (1992) *Plant Physiol.*, **98**, 573–577.

92 Lillo, C., Lea, U.S., Leydecker, M.T., and Meyer, C. (2003) *Plant J.*, **35**, 566–573.

93 Lea, U.S., Ten Hoopen, F., Provan, F., Kaiser, W.M., Meyer, C., and Lillo, C. (2004) *Planta*, **219**, 59–65.

94 Salgado, I., Modolo, L.V., Augusto, O., Braga, M.R., and Oliveira, H.C. (2007) Mitochondrial nitric oxide synthesis during plant–pathogen interactions: role of nitrate reductase in providing substrates, in *Nitric Oxide in Plant Growth, Development and Stress Physiology* (eds L. Lamattina and J.C. Polacco), Springer-Verlag, Berlin, pp. 239–254.

95 Oliveira, H.C., Justino, G.C., Sodek, L., and Salgado, I. (2009) *Plant Sci.*, **176**, 105–111.

96 Wang, R., Xing, X., and Crawford, N. (2007) *Plant Physiol.*, **145**, 1735–1745.

8
Effective Plant Protection Weapons Against Pathogens Require "NO Bullets"

Luzia V. Modolo

Summary

The mechanism by which plants defend themselves against pathogens must be highly efficient since plants cannot physically "escape" the threat. Oxidative burst and phytoalexin production are some initial events orchestrated in plant response. Nitric oxide has emerged as both a signaling and an exterminator molecule in plant defense. NO is a reactive nitrogen species that freely diffuses through membranes. It is widely produced by plant tissues and its origin appears to be much more complex than that in animals. The way by which NO contributes to plant defense has intrigued plant biologists for over a decade. How can such a minute molecule possess such power over pathogens? The functions of this versatile molecule in plant–pathogen interactions are discussed in detail in this chapter.

8.1
Introduction

In contrast to vertebrates, plants cannot physically flee their enemies; they must stand and fight. Plants may be targeted by herbivores or threatened by microbial pathogens (e.g., fungi, bacteria, viruses, among others), organisms possessing an array of effectors that facilitate host colonization. Here, our focus is on how plants defend themselves when potential pathogens attempt to infect them and disseminate. This "static defense" requires a highly concerted mechanism involving biochemical and molecular communications capable of preventing the spread of the invader.

The roles of nitric oxide in plants have been widely explored [1–10] since Lowell Klepper's discoveries demonstrated that herbicide-treated soybean plants emitted NO [11]. As a versatile free radical that diffuses into cell membranes, NO has

Nitric Oxide in Plant Physiology. Edited by S. Hayat, M. Mori, J. Pichtel, and A. Ahmad

ISBN: 978-3-527-32519-1

emerged as a key molecule in plant response to pathogens. Doke [12], Lamb [13], and Klessig's [14] groups were pioneers in this field, demonstrating the importance of NO in various pathways triggered during plant defense. By treating potato tuber tissue with the NO donor NOC-18, Noritake *et al.* observed the accumulation of rishitin (phytoalexin), an event prevented by the simultaneous application of NOC-18 and the NO scavenger cPTIO [12]. Delledonne *et al.* demonstrated that exogenous NO potentiated the hypersensitive response (HR) in soybean cell suspension cultures challenged with *Pseudomonas syringae* pv. *glycinea* [13]. In addition, suppression of endogenous NO by treatment of cell cultures with cPTIO compromised HR cell death, and the use of mammalian nitric oxide synthase (NOS) inhibitors provoked the reduction of *PAL* and *GST* transcription levels [13]. HR cell death in *Arabidopsis thaliana* was also compromised by mammalian NOS inhibitors [13]. Likewise, Durner *et al.* demonstrated that cyclic guanosine monophosphate (cGMP) and cyclic adenosine diphosphate ribose (cADPR) may function as second messengers for NO signaling as it occurs in mammals [14]. NO signaling through cGMP and cADPR was responsible for inducing the phenylalanine ammonia-lyase (*PAL*) and pathogenesis-related (*PR*) gene expression in tobacco resistant plants challenged with tobacco mosaic virus. Inhibitors of cGMP partially blocked the expression of the *PAL* gene, suggesting that an NO signaling pathway independent of cGMP was also involved in this event [14].

By using a series of different approaches, other groups have contributed to the elucidation of NO roles in plant–pathogen interactions [15–25]. The defense machinery of plants against microbial pathogens appears to be, in part, similar to that of animals. However, many differences exist, and perhaps what we know about NO signaling in plant–pathogen interactions is only the "tip" of a proverbial "iceberg," with many more discoveries to arise. The events that enable plants to defend themselves against microbial pathogens are interconnected, with most occurring simultaneously (Figure 8.1). Here, the roles of NO in such events are didactically separated into topics to facilitate the understanding of the complex network signaling pathways involved in plant response to pathogens.

8.2
Nitric Oxide and Reactive Oxygen Species in the Hypersensitive Response

An oxidative burst is one of the earliest events that occur during plant response to pathogen attack. The generation of reactive oxygen (e.g., O_2^- and H_2O_2) and nitrogen species (e.g., NO) is crucial for cell signaling and hypersensitive response, in which a limited number of plant cells in direct contact with the invading pathogen die rapidly to prevent proliferation of the invading organisms.

Reactive oxygen and nitrogen species (ROS and RNS, respectively) are likely produced temporally and spatially together and may directly kill pathogens.

As in animals, a neutrophil-like membrane-bound NADPH oxidase system is identified as the primary source of ROS during plant–pathogen interaction [26–28], but other sources, such as apoplastic peroxidases, and amine and

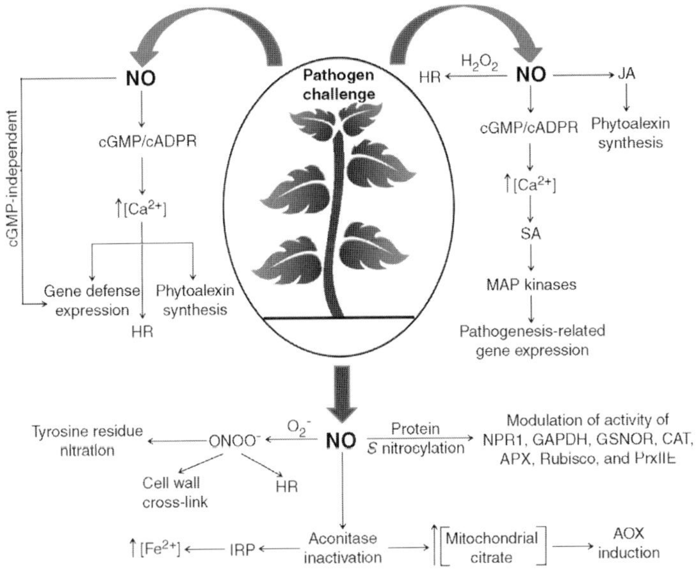

Figure 8.1 Roles of nitric oxide in plant defense against microbial pathogens. Events triggered by NO during plant response are didactically shown in separate schemes, although they are actually interconnected. AOX, alternative oxidase; APX, ascorbate peroxidase; cADPR, cyclic adenosine diphosphate ribose; CAT, catalase; cGMP, cyclic guanosine monophosphate; GADPH, glyceraldehyde 3-phosphate dehydrogenase; GSNOR, *S*-nitrosoglutathione reductase; HR, hypersensitive response; IRP, iron regulatory protein; JA, jasmonic acid; MAP kinases, mitogen-activated protein kinases; NPR1, *A. thaliana* protein that regulates salicylic acid-dependent gene expression; PrxIIE, peroxiredoxin II E; Rubisco, ribulose 1,5-bisphosphate carboxylase/oxygenase; SA, salicylic acid [29].

oxalate oxidase activities are described to contribute to some extent to ROS production [29].

The origin of NO in plant–pathogen interactions is still in debate. Enhanced AtNOA1 (nitric oxide-associated 1, an NOS-like enzyme) activity accounts for the massive production of NO in *A. thaliana* elicited with lipopolysaccharides from the surface of Gram-negative bacterial cells; similarly, mutant plants defective for *AtNOA1* gene failed to defend against the virulent *P. syringae* pv. *tomato* DC3000 [19]. In contrast, the NO burst in *A. thaliana* during interaction with *P. syringae* pv. *maculicola* appears to originate primarily from an NO_2^--reducing activity, dependent on complex III of the mitochondrial electron transport chain. The contribution of an NOS-like enzyme to the NO burst in this system is modest but likely sufficient for signaling purposes [21]. Furthermore, *A. thaliana* double mutants defective for nitrate reductase enzyme exhibited impaired HR when challenged with *P. syringae* pv. *maculicola*, due to the lack of the primary substrate (NO_2^-) for NO synthesis and also low arginine levels [24].

Regardless of its source in plants upon interaction with pathogens, NO, together with ROS, plays an important role in the signaling pathway that promotes HR cell death. Treatment of tobacco epidermal peels with elicitors to mimic a pathogen attack led to the production of ROS and NO with comparable time courses [30, 31]. The concomitant production of ROS and NO appears to be balanced [32], but the molecular mechanism by which ROS and NO simultaneously induce HR is not fully understood. NO can react with O_2^- to form a stronger oxidant species, the peroxinitrite anion ($ONOO^-$). The role of $ONOO^-$ in the *A. thaliana–P. syringae* interaction has been investigated using urate, a natural $ONOO^-$ scavenger [33]. Urate attenuated peroxidase activity in *A. thaliana* leaves challenged with *P. syringae* pv. *phaseolicola* (*avr*RPM1$^+$). The activity of peroxidase is known to increase during incompatible plant–pathogen interactions to promote cell wall cross-link. This suggests a role for $ONOO^-$ in HR-mediated alterations of the cell wall.

The chemical reaction between NO and H_2O_2 yields singlet oxygen ($^1O_2^*$) and hydroxyl radical ($^{\bullet}OH$), oxidant species that can cause cell death. Nevertheless, the relevance of this reaction to plant cell death during pathogen attack is not clear and demands further investigation.

Overexpression of a cysteine protease inhibitor in *A. thaliana* cell suspensions or tobacco plants repressed NO/ROS-mediated cell death indicating that an enzymatic mechanism is activated to control HR [34].

The availability of NO during HR was critically modulated by the relative rates of O_2^- and H_2O_2 production in soybean cell suspensions challenged with *P. syringae* pv. *glycinea* (*avrA*$^+$) [32]. When O_2^- production is favored in relation to that of NO, the latter is scavenged before it can react with H_2O_2. If the balance favors NO, O_2^- is rapidly scavenged instead of dismutating into H_2O_2. In addition, the cooperation between NO and H_2O_2 leads to cell death, while balanced amounts of NO and O_2^- result in the generation of $ONOO^-$. The peroxinitrite anion appears to be harmless to soybean cells since it is not an essential intermediate of NO-triggered cell death in this system [32]. On the other hand, studies performed with yeast elicitor-treated *Cupressus lusitanica* cell cultures reveal that NO and O_2^-, but not H_2O_2 itself, are required to induce cell death, highlighting the cytotoxicity of $ONOO^-$ in this biological system [35]. The induction of gene expression of pathways that lead to the production of antimicrobial compounds may also be taken into account when analyzing the divergent role of $ONOO^-$ in the death of elicitor-challenged *C. lusitanica* and soybean cells. Different pathways for the biosynthesis of antimicrobial compounds were activated in soybean and *C. lusitanica* cells (flavonoids in the former and tropolone in the latter) to the extent that is not reported. Thus, a differential contribution of antimicrobial compounds to cell death may be occurring in both cell suspension cultures in addition to the $ONOO^-$ contribution.

NO, in conjunction with H_2O_2, modulates the gene expression of enzymes that are responsive to oxidative stress (glutathione peroxidase and glutathione *S*-transferase) [10, 14]. NO and H_2O_2 also induce cytochrome *c* release from the mitochondrial membrane during HR cell death [36].

NO-triggered cell death in tobacco and *A. thaliana* apparently requires cGMP and/or cyclic ADPR as second messengers [14, 15]. Cyclic GMP is a second messenger in animals that can regulate directly or indirectly many physiological functions by activating different types of calcium channels. A similar action may take place in plants since *A. thaliana* plants defective for a nucleotide-gated channel (AtCNGC2) were unable to develop HR [37, 38]. Indeed, NO accumulation in cryptogenin-elicited tobacco cells depends on calcium influx from extracellular space. Once inside the cell, NO mediates calcium mobilization, a signal that leads the cells to undergo death [39]. Calcium channel inhibitors also prevent NO accumulation in oligogalacturonic acid-elicited ginseng suspension cultures [40].

A gain-of-function approach has been used to confirm that NO is a signaling molecule during HR processes. Zeier *et al.* overexpressed a bacterial *NOD* gene in *A. thaliana* plants [41]. This gene encodes a dioxygenase enzyme that catalyzes NO oxidation to NO_3^-. Transgenic plants challenged with an avirulent strain of *P. syringae* pv. tomato (avrB^+) exhibited low NO levels and compromised HR [41], reinforcing the conclusion that NO serves as a signaling molecule to induce HR.

8.3 Nitric Oxide and Phytoalexin Production

Production of antimicrobial compounds termed phytoalexins is also part of the defense mechanism developed by plants to avoid pathogen invasion and dissemination [42]. The first evidence showing that NO induced phytoalexin accumulation originated from the use of NO donors and/or the NO scavenger cPTIO [12]. Potato tuber tissues accumulated the phytoalexin rishitin after exposure to the NO donor NOC-18, a process inhibited by an NO scavenger [12]. NO-dependent accumulation of hypericin was observed in *Hypericum perforatum* cell suspensions treated with elicitors from *Aspergillus niger* cell wall [43].

Transcription levels of *PAL* gene that encodes phenylalanine ammonia-lyase, the first enzyme of the phenylpropanoid pathway, increased during plant–pathogen interaction (when NO levels are increased), leading to the accumulation of phytoalexins [13]. NO also stimulated an increase in the transcription levels of genes that encode other enzymes of the phenylpropanoid pathway such as cinnamic acid-4-hydroxylase (C4H) and chalcone synthase (CHS) [13].

The involvement of an NOS-like enzyme in the generation of NO has been demonstrated in plant–pathogen interactions where phytoalexin production was investigated. The first report was published by Modolo *et al.* in 2002 [17]. The authors found that an NOS-like enzyme reaches its maximum activity prior to the accumulation of phytoalexins in soybean cotyledons treated with oligosaccharides from *Diaporthe phaseolorum* pv. *meridionalis* [17]. The phytoalexin synthesis in elicited soybean cotyledons was partially inhibited by mammalian NOS inhibitors (L-NAME or AMG) suggesting that other sources of NO or other molecules are implicated in the signaling process. Accumulation of the tropolone phytoalexin β-thujaplicin in yeast elicitor-treated *C. lusitanica* cells is partially blocked by mammalian NOS inhibitors [35]. Increased production of the antimicrobial saponin in oligogalacturonic acid-elicited ginseng cell cultures is also accompanied by an increased NOS-like enzyme activity [40].

8.4
Nitric Oxide and the Salicylic Acid Signaling Pathway

Salicylic acid (SA) is an important signaling molecule in plant defense that triggers the accumulation of pathogenesis-related proteins, leading to local and systemic disease resistance. This section focuses on the role of NO in the salicylic acid signaling pathway. Detailed information on the interplay between nitric oxide and other signals in plant resistance to pathogens is provided in Chapter 11.

Activation of mitogen-activated protein kinases (MAPKs) by NO was reported in tobacco and *A. thaliana* plants [14–16]. NO activates tobacco SIPK (salicylic acid-induced protein kinase); such activation depends on SA since transgenic NahG tobacco plants defective for SA accumulation experience suppressed SIPK activation in leaves injected with recombinant NOS enzyme [16]. Induction of pathogenesis-related protein 1 (PR-1) gene expression by NO is also repressed in transgenic NahG tobacco plants indicating that SA is a mediator in this process and its production depends on NO [14].

Cytosolic and mitochondrial aconitase enzymes are also targeted by NO yielding inactivated proteins [44]. The mechanism of aconitase inactivation by NO is still unclear, however. Inhibition of cytosolic aconitase may trigger its conversion to an iron regulatory protein (IRP) as occurs in animals, which facilitates an increase in intracellular free Fe^{2+}. In tobacco, repression of mitochondrial aconitase activity may function as a defense against tobacco mosaic virus and other viruses by promoting citrate accumulation in mitochondria that leads to induction of alternative oxidase (AOX) [45]. The effects of NO on aconitase (and also on ascorbate peroxidase (APX) and catalase) may be mediated by SA [46].

An *S*-nitrosoglutathione reductase enzyme (GSNOR) was recently reported to play an important role in *S*-nitrosothiol protein homeostasis in *A. thaliana* [20]. Over-expression of *AtGSNOR1* gene in *A. thaliana* plants promoted a dramatic reduction in *S*-nitrosothiol protein levels and enhanced plant defense against virulent microbial pathogens. AtGSNOR1 activity led to the formation of free NO, establishing *S*-nitrosothiol protein turnover. Interestingly, the SA signaling pathway was positively regulated by AtGSNOR1 protein [20].

S-Nitrosylation of *A. thaliana* NPR1 protein was recently found to be critical for its function due to changes in protein conformation [47]. NPR1 is essential for regulating SA-dependent gene expression during *A. thaliana* defense. NPR1 is located in the cytoplasm of unchallenged plant cells as an oligomer that is maintained by intermolecular disulfide bonds. The increase in SA levels during pathogen challenge promotes changes in the cell redox state with consequent reduction of disulfide bonds in NPR1 structure. Monomeric units are then released and translocated to the nucleus to activate the expression of *PR* genes. The NO donor nitrosoglutathione (GSNO), but not sodium nitroprusside (SNP), facilitates NPR1 oligomerization by *S*-nitrosylating, at least the residue Cys156 [47]. *S*-Nitrosylation of NPR1 appears to be responsible for the maintenance of NPR1 homeostasis upon SA activation. This effect was attested by results obtained with transgenic *npr1 A. thaliana* plants harboring *NPR1-C156A-GFP* mutant gene. These plants experienced protein nuclear localization and enhanced resistance to *P. syringae* pv. *maculicola*. However, NPR1-C156A-GFP protein was depleted 48 h after plant treatment with SA, compromising pathogen resistance [47].

Considering the different effects of GSNO and SNP on this system, further studies are necessary to identify which NO species (NO^+, NO^-, or $NO^{\bullet}$) are actually contributing to NPR1 oligomerization.

8.5
Nitric Oxide and the Jasmonic Acid Signaling Pathway

Jasmonic acid (JA) is a fatty acid-derived signaling molecule involved in a variety of biological processes including plant defense against pathogens [48]. Various *A. thaliana* mutant plants impaired in jasmonic acid production exhibited enhanced susceptibility to both fungal and bacterial pathogens [49–51]. Treatment of SA-deficient plants (*NahG* mutants) with NO induced an increase in JA levels and subsequent increase in *PDF-1.2* and *JIP* gene expression [10], indicating that the NO-dependent JA production is negatively modulated by SA. Biosynthesis of NO, JA, and the phytoalexin hypericin was induced in *H. perforatum* cell suspensions upon treatment with fungal elicitor from *A. niger* cell wall [43]. Inhibitors of JA biosynthesis do not prevent NO formation, suggesting that JA acts downstream of the NO signaling pathway. Concomitant use of NO scavenger and JA biosynthesis inhibitor blocks the elicitor-induced accumulation of hypericin in *H. perforatum* cell suspensions indicating that NO signaling in this cell system depends on JA.

8.6
Nitric Oxide and Gene Regulation

Pharmacological studies were used to address the transcriptional changes that occur in *A. thaliana* and tobacco plants as a result of NO and/or H_2O_2 signaling [41, 52, 53]. Signal transduction-related genes (e.g., transcription factors *WRKY*), mitogen-activated protein kinases, and genes involved in plant defense (e.g., *GST, GRX*, and *CAT*) were modulated by NO. An overlap between genes regulated by H_2O_2 or NO occurred in tobacco plants infiltrated with the NO donor SNP, indicating an interplay between these signaling molecules [54]. Different NO donors, however, provide different patterns of transcriptional changes as observed by Ferrarini *et al.* in studies conducted with the legume *Medicago truncatula* [25]. Only 11% of the genes up- or down-regulated in leaves by at least twofold suffered the same modulatory effect by GSNO and SNP; this number dropped to 1.6% in root tissues [25]. Thus, NO redox state is critical for gene expression regulation.

The interaction between *A. thaliana* and *Colletotrichum trifolii* race 1 induces high accumulation of NO with a range of responsive genes. Although this approach does not permit the conclusion that such genes are directly responsive to NO, the results obtained from application of NO donors, in association with those from plant–pathogen interaction, open a new window for further investigations into the role of NO in *M. truncatula* defense.

Treatment of *A. thaliana* cell suspensions with the NO donor NOR-3 promoted the induction of several genes, among them *AOX1a*, which encodes an alternative oxidase [55]. AOX proteins disrupt the mitochondrial electron transport chain by receiving electrons from ubiquinol in a nonphosphorylative process that results in energy release in the form of heat. The expression of *AOX1a* gene in *A. thaliana* is induced not only by exogenous NO but also by the interaction with either *P. syringae* pv. tomato or proteinaceous bacterial elicitor Harpin [56, 57].

8.7
Nitric Oxide and Protein Regulation

Proteins other than those involved in the SA signaling pathway (see Section 8.4) are regulated by NO. Participation of NO in redox signaling during plant–pathogen interaction is also evidenced by its reversible inhibitory effect on tobacco catalase and ascorbate peroxidase, enzymes involved in H_2O_2 scavenging [5, 58–60]. Inhibition of CAT and APX activities allows the accumulation of H_2O_2 and potentiates the effect of ROS during HR cell death. However, it is not known how NO affects the activities of these enzymes. The inhibitory effect of NO on such enzymes is also observed in plant–necrotrophic pathogen interactions where a mechanism other than *R*-gene resistance occurs [61, 62].

Protein *S*-nitrosylation is a posttranslational modification by which NO (mostly as nitrosonium cation, NO^+) modulates the activity of certain enzymes, especially those related to the plant antioxidant system. The peroxidase and peroxinitrite reductase activities of the peroxiredoxin II E (PrxIIE) enzyme were inhibited by enzyme *S*-nitrosylation and this inhibition led to the accumulation of $ONOO^-$ and consequent protein tyrosine nitration [63]. Tyrosine kinase signaling is affected by nitration of tyrosine residues. In addition, nitration of tyrosine residues may either alter protein conformation and activity or make the protein a target of proteases [64]. Glyceraldehyde 3-phosphate dehydrogenase (GAPDH) is another example of a protein that is inhibited by *S*-nitrosylation of the cysteine(s) sulfhydryl group [65]. GAPDH plays an important role in glycolysis, but it may have an additional function in mediating ROS signaling during plant–pathogen interactions [66]. Recently, the role of NO in regulating ribulose 1,5-bisphosphate carboxylase/oxygenase (Rubisco) activity was reported [67]. This enzyme acts in the Calvin Cycle, catalyzing the first step of carbon fixation. Rubisco is capable of catalyzing either the carboxylation or oxygenation of ribulose 1,5-bisphosphate. By using MALDI–TOF mass spectrometry, Abat *et al.* uncovered the *S*-nitrosoproteome of a crassulacean acid metabolism in *Kalanchoe pinnata*. *S*-Nitrosylation of cysteine residue present in the small subunit of

Rubisco protein inhibited carbon fixation [67]. The mRNA and protein levels of Rubisco are known to be reduced in potato plants under pathogen challenge or elicitor treatment [68]. Thus, *S*-nitrosylation promoted by NO may serve as an additional mechanism to repress Rubisco activity as part of the plant defense mechanism against pathogens.

8.8
Concluding Remarks

The efficiency of the plant defense mechanism dictates whether or not microbial pathogens will succeed in their attempt to establish infection. An increasing body of evidence identifies NO as a critical signaling molecule for plant response to pathogens. A likely cross talk between NO and ROS, SA, or JA signaling pathways reveals the vast arsenal used by plants for self-protection. We may be far from fully elucidating the roles NO plays in plant defense, but the amount of information known to date indicates a sound beginning. The use of transgenic plants has helped identify certain molecules that work downstream of the NO signaling pathway. However, gain- and loss-of-function approaches have much more to contribute to our understanding of how NO fits into the larger puzzle that drives plant defense.

From all these discoveries, it is noteworthy that the weapon used by plants against pathogens requires "NO bullets."

Acknowledgment

The author is grateful to Dr. Ângelo de Fátima for the artwork.

References

1 Millar, A.H. and Day, D.A. (1996) *FEBS Lett.*, **398**, 155–158.
2 Keeley, J.E. and Fotheringham, C.J. (1997) *Science*, **276**, 1248–1250.
3 Beligni, M.V. and Lamattina, L. (1997) *Nitric Oxide*, **3**, 199–208.
4 Ribeiro, E.A., Jr., Cunha, F.Q., Tamashiro, W.M., and Martins, I.S. (1999) *FEBS Lett.*, **445**, 283–286.
5 Clark, D., Durner, J., Navarre, D.A., and Klessig, D.F. (2000) *Mol. Plant–Microbe Interact.*, **13**, 1380–1384.
6 Mata, C.G. and Lamattina, L. (2001) *Plant Physiol.*, **126**, 1196–1204.
7 Yamasaki, H., Shimoji, H., Ohshiro, Y., and Sakihama, Y. (2001) *Nitric Oxide*, **5**, 261–270.
8 Zottini, M., Formentin, E., Scattolin, M., Carimi, F., Lo Schiavo, F., and Terzi, M. (2002) *FEBS Lett.*, **515**, 75–78.
9 Dordas, C., Rivoal, J., and Hill, R.D. (2003) *Ann. Bot.*, **91**, 173–178.
10 Grun, S., Lindermayr, C., Sell, S., and Durner, J. (2006) *J. Exp. Bot.*, **57**, 507–516.
11 Klepper, L. (1979) *Atmospheric Environment*, **13**, 537–542.
12 Noritake, T., Kawakita, K., and Doke, N. (1996) *Plant Cell Physiol.*, **37**, 113–116.

13 Delledonne, M., Xia, Y., Dixon, R.A., and Lamb, C. (1998) *Nature*, **394**, 585–588.
14 Durner, J., Wendehenne, D., and Klessig, D.F. (1998) *Proc. Natl. Acad. Sci. USA*, **95**, 10328–10333.
15 Clarke, A., Desikan, R., Hurst, R.D., Hancock, J.T., and Neill, S.J. (2000) *Plant J.*, **24**, 667–677.
16 Kumar, D. and Klessig, D.F. (2000) *Mol. Plant–Microbe Interact.*, **13**, 347–351.
17 Modolo, L.V., Cunha, F.Q., Braga, M.R., and Salgado, I. (2002) *Plant Physiol.*, **130**, 1288–1297.
18 Delledonne, M., Polverari, A., and Murgia, I. (2003) *Antioxid. Redox. Signal.*, **5**, 33–41.
19 Zeidler, D., Zahringer, U., Gerber, I., Dubery, I., Hartung, T., Bors, W., Hutzler, P., and Durner, J. (2004) *Proc. Natl. Acad. Sci. USA*, **101**, 15811–15816.
20 Feechan, A., Kwon, E., Yun, B.W., Wang, Y., Pallas, J.A., and Loake, G.J. (2005) *Proc. Natl. Acad. Sci. USA*, **102**, 8054–8059.
21 Modolo, L.V., Augusto, O., Almeida, I.M., Magalhaes, J.R., and Salgado, I. (2005) *FEBS Lett.*, **579**, 3814–3820.
22 Mur, L.A., Santosa, I.E., Laarhoven, L.J., Holton, N.J., Harren, F.J., and Smith, A.R. (2005) *Plant Physiol.*, **138**, 1247–1258.
23 Sokolovski, S., Hills, A., Gay, R., Garcia-Mata, C., Lamattina, L., and Blatt, M.R. (2005) *Plant J.*, **43**, 520–529.
24 Modolo, L.V., Augusto, O., Almeida, I.M.G., Pinto-Maglio, C.A.F., Oliveira, H.C., Seligman, K., and Salgado, I. (2006) *Plant Sci.*, **171**, 34–40.
25 Ferrarini, A., De Stefano, M., Baudouin, E., Pucciariello, C., Polverari, A., Puppo, A., and Delledonne, M. (2008) *Mol. Plant–Microbe Interact.*, **21**, 781–790.
26 Keller, T., Damude, H.G., Werner, D., Doerner, P., Dixon, R.A., and Lamb, C. (1998) *Plant Cell*, **10**, 255–266.
27 Torres, M.A., Onouchi, H., Hamada, S., Machida, C., Hammond-Kosack, K.E., and Jones, J.D. (1998) *Plant J.*, **14**, 365–370.
28 Torres, M.A., Dangl, J.L., and Jones, J.D. (2002) *Proc. Natl. Acad. Sci. USA*, **99**, 517–522.
29 Bolwell, G.P. and Wojtaszek, P. (1997) *Physiol. Mol. Plant Pathol.*, **51**, 347–366.
30 Allan, A.C. and Fluhr, R. (1997) *Plant Cell*, **9**, 1559–1572.
31 Foissner, I., Wendehenne, D., Langebartels, C., and Durner, J. (2000) *Plant J.*, **23**, 817–824.
32 Delledonne, M., Zeier, J., Marocco, A., and Lamb, C. (2001) *Proc. Natl. Acad. Sci. USA*, **98**, 13454–13459.
33 Alamillo, J.M. and Garcia-Olmedo, F. (2001) *Plant J.*, **25**, 529–540.
34 Belenghi, B., Acconcia, F., Trovato, M., Perazzolli, M., Bocedi, A., Polticelli, F., Ascenzi, P., and Delledonne, M. (2003) *Eur. J. Biochem.*, **270**, 2593–2604.
35 Zhao, J., Fujita, K., and Sakai, K. (2007) *New Phytol.*, **175**, 215–229.
36 Mur, L.A., Carver, T.L., and Prats, E. (2006) *J. Exp. Bot.*, **57**, 489–505.
37 Leng, Q., Mercier, R.W., Yao, W., and Berkowitz, G.A. (1999) *Plant Physiol.*, **121**, 753–761.
38 Clough, S.J., Fengler, K.A., Yu, I.C., Lippok, B., Smith, R.K.J., and Bent, A.F. (2000) *Proc. Natl. Acad. Sci. USA*, **97**, 9323–9328.
39 Lamotte, O., Gould, K., Lecourieux, D., Sequeira-Legrand, A., Lebrun-Garcia, A., Durner, J., Pugin, A., and Wendehenne, D. (2004) *Plant Physiol.*, **135**, 516–529.
40 Hu, X.Y., Neill, S.J., Cai, W.M., and Tang, Z.C. (2003) *Funct. Plant Biol.*, **30**, 901–907.
41 Zeier, J., Delledonne, M., Mishina, T., Severi, E., Sonoda, M., and Lamb, C. (2004) *Plant Physiol.*, **136**, 2875–2886.
42 de Fátima, A. and Modolo, L.V. (2008) *Advances in Plant Biotechnology* (eds G.P. Rao, Y. Zhao, V.V. Radchuk, and S.K. Bhatnagar), Studium Press LLC, Houston, pp. 179–204.
43 Xu, M.J., Dong, J.F., and Zhu, M.Y. (2005) *Plant Physiol.*, **139**, 991–998.
44 Navarre, D.A., Wendehenne, D., Durner, J., Noad, R., and Klessig, D.F. (2000) *Plant Physiol.*, **122**, 573–582.
45 Vanlerberghe, G.C. and McIntosh, L. (1996) *Plant Physiol.*, **111**, 589–595.
46 Klessig, D.F., Durner, J., Noad, R., Navarre, D.A., Wendehenne, D., Kumar, D., Zhou,

J.M., Shah, J., Zhang, S., Kachroo, P., Trifa, Y., Pontier, D., Lam, E., and Silva, H. (2000) *Proc. Natl. Acad. Sci. USA*, **97**, 8849–8855.
47 Tada, Y., Spoel, S.H., Pajerowska-Mukhtar, K., Mou, Z., Song, J., Wang, C., Zuo, J., and Dong, X. (2008) *Science*, **321**, 952–956.
48 Reymond, P. and Farmer, E.E. (1998) *Curr. Opin. Plant. Biol.*, **1**, 404–411.
49 Staswick, P.E., Yuen, G.Y., and Lehman, C.C. (1998) *Plant J.*, **15**, 747–754.
50 Thomma, B.P., Eggermont, K., Penninckx, I.A., Mauch-Mani, B., Vogelsang, R., Cammue, B.P., and Broekaert, W.F. (1998) *Proc. Natl. Acad. Sci. USA*, **95**, 15107–15111.
51 Stintzi, A., Weber, H., Reymond, P., Browse, J., and Farmer, E.E. (2001) *Proc. Natl. Acad. Sci. USA*, **98**, 12837–12842.
52 Polverari, A., Molesini, B., Pezzotti, M., Buonario, R., Marte, M., and Delledonne, M. (2003) *Mol. Plant–Microbe Interact.*, **16**, 1094–1105.
53 Parani, M., Rudrabhatla, S., Myers, R., Weirich, H., Smith, B., Leaman, D.W., and Goldman, S.L. (2004) *Plant Biotechnol. J.*, **2**, 359–366.
54 Zago, E., Morsa, S., Dat, J.F., Alard, P., Ferrarini, A., Inze, D., Delledonne, M., and Van Breusegem, F. (2006) *Plant Physiol.*, **141**, 404–411.
55 Huang, X., von Rad, U., and Durner, J. (2002) *Planta*, **215**, 914–923.
56 Simons, B.H., Millenaar, F.F., Mulder, L., Van Loon, L.C., and Lambers, H. (1999) *Plant Physiol.*, **120**, 529–538.
57 Krause, M. and Durner, J. (2004) *Mol. Plant–Microbe Interact.*, **17**, 131–139.
58 Bestwick, C.S., Adam, A.L., Puri, N., and Mansfield, J.W. (2001) *Plant Sci.*, **161**, 497–506.
59 de Pinto, M.C., Tommasi, F., and De Gara, L. (2002) *Plant Physiol.*, **130**, 698–708.
60 Mittler, R., Lam, E., Shulaev, V., and Cohen, M. (1999) *Plant Mol. Biol.*, **39**, 1025–1035.
61 Floryszak-Wieczorek, J., Milczarek, G., Arasimowicz, M., and Ciszewski, A. (2006) *Planta*, **224**, 1363–1372.
62 Floryszak-Wieczorek, J., Arasimowicz, M., Milczarek, G., Jelen, H., and Jackowiak, H. (2007) *New Phytol.*, **175**, 718–730.
63 Romero-Puertas, M.C., Laxa, M., Matte, A., Zaninotto, F., Finkemeier, I., Jones, A.M.E., Perazzolli, M., Vandelle, E., Dietz, K.J., and Delledonne, M. (2007) *Plant Cell*, **19**, 4120–4130.
64 Souza, J.M., Choi, I., Chen, Q., Weisse, M., Daikhin, E., Yudkoff, M., Obin, M., Ara, J., Horwitz, J., and Ischiropoulos, H. (2000) *Arch. Biochem. Biophys.*, **380**, 360–366.
65 Lindermayr, C., Saalbach, G., and Durner, J. (2005) *Plant Physiol.*, **137**, 921–930.
66 Hancock, J.T., Henson, D., Nyirenda, M., Desikan, R., Harrison, J., Lewis, M., Hughes, J., and Neill, S.J. (2005) *Plant Physiol. Biochem.*, **43**, 828–835.
67 Abat, J.K., Mattoo, A.K., and Deswal, R. (2008) *FEBS J.*, **275**, 2862–2872.
68 Logemann, E., Wu, S.C., Schroder, J., Schmelzer, E., Somssich, I.E., and Hahlbrock, K. (1995) *Plant J.*, **8**, 865–876.

9
The Role of Nitric Oxide as a Bioactive Signaling Molecule in Plants Under Abiotic Stress

Gang-Ping Hao and Jian-Hua Zhang

Summary

Nitric oxide is a short-lived bioactive molecule initially described as a toxic compound but now recognized as an important signaling and effector molecule both in animal and in plant cells. NO is proposed to be one of the important second (should be second) messengers in plant cells. Various data indicate that NO is an endogenous signal in plants that mediates responses to several stimuli. Experimental evidence in support of such a signaling role of NO has been obtained through the application of NO, usually in the form of NO donors, through the measurement of endogenous NO, and through the manipulation of endogenous NO content by chemical and genetic means. In the last few years, many studies have described NO as both a cytotoxic and a cytoprotecting regulator involved in different physiological responses to abiotic stress in plants such as drought, UV-B, salinity, and high temperature. For example, application of the NO donor, sodium nitroprusside, confers resistance to salt, drought, heavy metals, and chilling stresses. In plants, the sources of NO production have been the subject of considerable debate. A growing body of evidence indicates that NO is formed by mammalian-like NOS activity, nitrate reductase, or nonenzymatic sources. New evidence involving NO in signal transduction pathways mediated by some key molecules such as cyclic guanosine monophosphate, cyclic adenosine diphosphate ribose, and Ca^{2+} recently have been reported in plants. There is also a compelling evidence suggesting that abscisic acid, hydrogen peroxide, and NO interact under abiotic stress in plants. In this chapter, nitric oxide functions as a bioactive signaling molecule in plant abiotic stress responses are discussed. The cross talk between NO and other key signaling components under abiotic stress is also reviewed.

Nitric Oxide in Plant Physiology. Edited by S. Hayat, M. Mori, J. Pichtel, and A. Ahmad

ISBN: 978-3-527-32519-1

9.1
Introduction

Nitric oxide (NO) is a short-lived bioactive molecule initially described as a toxic compound but now recognized as an important signaling and effector molecule both in animal and in plant cells [1]. Nitric oxide was first detected in mammals, where it plays various roles ranging from dilation of blood vessels to neurotransmission and immune response. In comparison to animal studies, relatively little is known about the biological functions of NO in plants. Earliest studies on the effect of NO on plants were concerned with the toxic action of nitric oxides (NO_2, N_2O_3, NO^{2-}, and NO^{3-}), primarily on the photosynthetic apparatus of plants and chlorophyll levels in selected forest tree species and plants in parks or industrial areas [2]. Groundbreaking studies showing NO as a signaling molecule in plants appeared as late as 1998 [3, 4] and as a consequence resulted in the intensification of research on the role of NO in plants. The last few years have seen an increasing number of studies dedicated to NO functions in plants. NO appears to be involved in plant development processes and participates in a number of physiological processes such as growth, pathogen defense response, development, programmed cell death, flowering, stomatal closure, and response to environmental stresses [5]. Although the understanding of mechanisms by which NO contributes to these processes is still in its infancy, promising results are being obtained.

9.2
Biosynthesis of Nitric Oxide Under Abiotic Stress

In plants, the sources of NO production have been the subject of much debate. A growing body of evidence indicates that NO is formed by mammalian-like nitric oxide synthase (NOS) activity, nitrate reductase (NR), as well as the nonenzymatic pathway in the apoplast [2]. Several potential NO sources may be distinguished in plants, with the physiological role of each depending on the species, type of tissue or cells, external conditions, and potential activation of the signal pathway.

9.2.1
NO Generated from NOS-Like Activity Under Abiotic Stress

In animal systems, most of the NO produced is due to the enzyme nitric oxide synthase (EC 1.14.13.3 9) [6, 7]. This enzyme catalyzes the oxygen- and NADPH-dependent oxidation of L-arginine to NO and citrulline in a complex reaction requiring FAD, FMN, tetrahydrobiopterin (BH_4), calcium, and calmodulin [8, 9]. Although NOS-like activity has been detected widely in plants, animal-type NOS is still elusive. Inhibitors of mammalian NOS were shown to inhibit NO generation in plants [2]. Moreover, the presence of NOS-like proteins in plant tissue was detected using immunoassays and immunocytochemical analyses with antibodies against animal NOS isoforms [2]. However, results of immunoassays remain dubious, as antibodies against mammalian NOS may recognize many plant proteins not connected with NOS [2]. Recently, in pea seedlings, using the chemiluminescence assay, Corpas *et al.* [10] demonstrated arginine-dependent NOS activity which was constitutive, sensitive to an irreversible inhibitor of animal NOS, and dependent on the plant organ and development stage.

A unique *Arabidopsis* (*Arabidopsis thaliana*) AtNOS1 gene that was suggested to encode a protein with sequence similarity to a protein that is involved in NO synthesis in the snail *Helix pomatia* has been isolated [11]. Atnos1 is a homozygous mutant line with T-DNA insertion in the first exon of NOS1 gene [11]. The *in vivo* NOS activity in the mutant is reduced to approximately 25% compared to that of wild type, and Atnos1 mutant has impaired NO production. However, the most recent studies have raised some critical questions regarding the nature of AtNOS1 [12, 13]. The findings that the recombinant AtNOS1 protein does not exhibit NOS activity *in vitro* suggest that the involvement of AtNOS1 in NO biosynthesis and accumulation may be either indirect or regulatory [12]. Therefore, AtNOS1 appears unlikely to be an Arg-dependent NOS enzyme; rather, it is likely that the AtNOS1 is a protein associated with the NO biosynthesis. Accordingly, the AtNOS1 was suggested to be renamed as AtNOA1 for NO-associated 1 [12]. Regardless of the nature of AtNOA1, the identification of AtNOA1 provides a powerful tool to control *in vivo* NOS activity as well as endogenous NO levels for determining the physiological function of NO, as the Atnoa1 mutants have impaired NOS activity and reduced endogenous NO levels [11, 14, 15]. Previous studies have shown that NOA1-dependent NO synthesis is involved in hormonal signaling, stomatal movement, flowering, pathogen defense, and oxidative stress [11, 14–16].

In the past few years, many studies have shown that NO dependent on NOS-like activity plays an important role in defense response to abiotic stress. It has been shown that NOS is localized in cytoplasm, chloroplasts, mitochondria, peroxisomes, and the nucleus in plant cells [17]. In a recent work at Jiang's laboratory [17], the activity of NOS in the cytosolic and microsomal fractions of maize leaves was determined. The results showed that water stress induced increases in NOS activity in the cytosolic and microsomal fractions, and the NOS activity in the microsomal fraction was higher and more susceptible to water stress treatment than that in the cytosolic fraction of maize leaves. Pretreatment with NOS inhibitors L-NAME (N^G-nitro-L-arginine methyl ester) and PBITU completely blocked increases in NOS activity in the cytosolic and microsomal fractions induced by water stress treatment. These results clearly suggest that water stress-induced increases in the production of NO are mainly from NOS. Another result from Jiang's laboratory [18] indicated that pretreatment with L-NAME substantially reduced the abscisic acid (ABA)-induced production of NO in maize leaves. These results suggest that NO dependence on NOS-like activity serves as a signaling component in the induction of protective responses and is associated with drought tolerance in maize seedlings. We investigated whether NO dependence on NOS activity is involved in the signaling of drought-induced protective responses in maize seedlings. The results showed that both NOS activity and the rate of NO release increased substantially under dehydration stress. The high NOS activity induced by c-PTIO(2-(4-carboxyphenyl)-4,4,5,5-

tetramethylimidazoline-1-oxyl-3-oxide) as NO scavenger, and NO accumulation inhibited by NOS inhibitor L-NAME in dehydration-treated maize seedlings indicated that most NO production under water deficit stress may be generated from NOS-like activity [19].

To characterize the role of NO in the tolerance of *Arabidopsis* (*A. thaliana*) to salt stress, the effect of NaCl on *Arabidopsis* wild-type and mutant (Atnoa1) plants with an impaired *in vivo* NOS activity and a reduced endogenous NO level was investigated. The results indicate that disruption of NOS-dependent NO production is closely related to the sensitivity of *Arabidopsis* to salt stress [20]. In another study, *Atnos1* mutant with decreased NO production due to a T-DNA insertion in *AtNOA1* gene was shown to be hypersensitive to oxidative stress induced by NaCl and MV [16]. In reed callus, NOS activity increased by 22% under NaCl treatment and N^{ω}-nitro-L-arginine (L-NNA) inhibited NOS activity. Inhibition of NO production by specific inhibitors such as L-NNA indicates that NO was produced from NOS-like activity [21].

NO generated by NOS-like activity is also induced by heat stress. For instance, exposure of zooxanthellae to heat stress (33 °C) for 24 h led to an increase in NOS-like activity [22]. Bouchard and Yamasaki [23] also reported that zooxanthellae from bleaching corals had higher NO production generated from NOS-like activities than zooxanthellae isolated from nonbleaching corals under 27–41 °C heat stress. Some experimental results also showed that UV radiation can induce the NOS activity increase and NO production enhancement. Mackerness *et al.* [24] reported the participation of NO from NOS-type enzymatic activity in plant response to UV-B radiation. The results of Qu *et al.* [25] showed that UV-B radiation significantly induced NOS activity and promoted NO release. Zhang *et al.* [26] found that NO produced from NOS-like activity was a second messenger associated with developmental growth under UV-B radiation. The results of Wang *et al.* [27] indicated that NO generated from NOS-like activity appeared to act in the same direction or synergistically with reactive oxygen species (ROS) to induce ethylene biosynthesis in defense response under UV-B radiation in leaves of maize seedlings.

In addition to drought, salt and UV-B radiation, other abiotic stresses such as heavy metal and mechanical stress also stimulate NO production through NOS-like activity. In *Arabidopsis*, mechanical stress induced NO production, and inhibition by L-NAME led Garcês *et al.* [28] to conclude that an inducible form of NOS was present. To elucidate the role of NO in Al toxicity to *Hibiscus moscheutos*, a herbaceous perennial horticultural plant, Tian *et al.* [29] investigated effects of Al on endogenous NO concentrations and activities of NOS and NR, and compared the effects with those of known NOS inhibitors, NO donors, and NO scavengers. The result showed that Al inhibited activity of NOS and reduced endogenous NO concentrations. The alleviation of inhibition of root elongation induced by Al, NO scavenger, and NOS inhibitor was correlated with endogenous NO concentrations in root apical cells (Figure 9.1), suggesting that reduction of endogenous NO concentrations resulting from inhibition of NOS activity could underpin Al-induced arrest of root elongation in *H. moscheutos*.

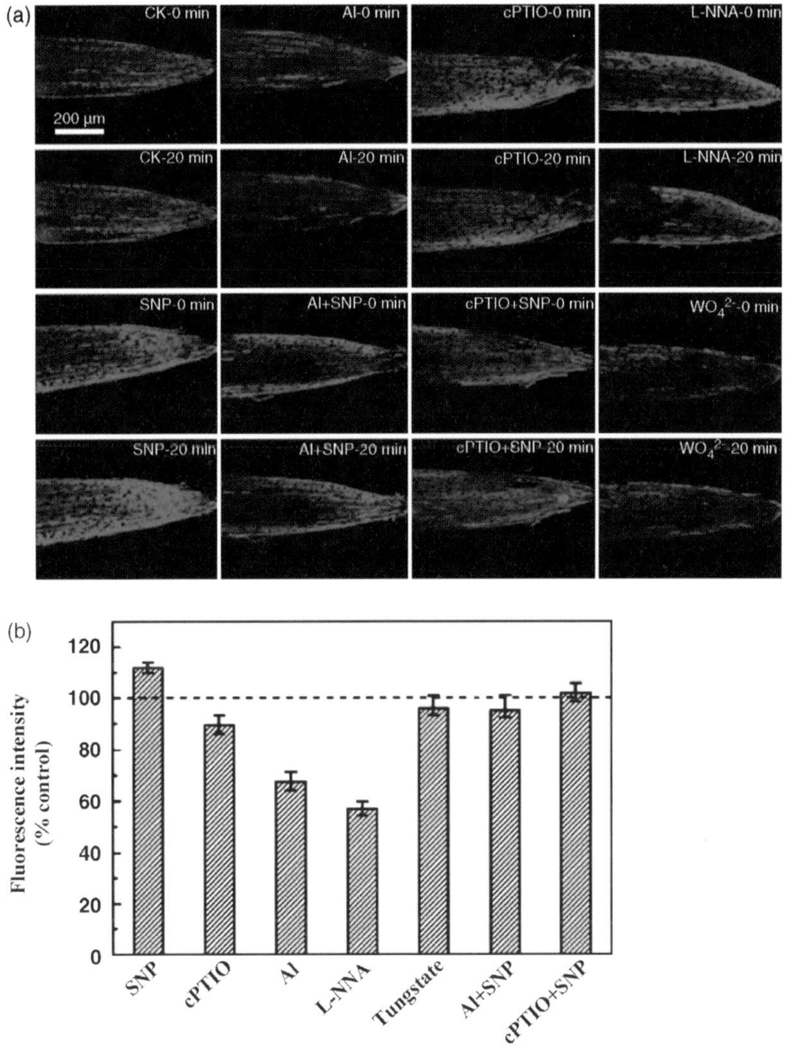

Figure 9.1 Effect of $AlCl_3$, N^{ω}-nitro-L-arginine (L-NNA), tungstate (Na_2WO_4), c-PTIO, and SNP on endogenous NO concentration in *H. moscheutos* roots. Roots of 4-day-old *H. moscheutos* were amended with 20 μM 4,5-diaminofluorescence (DAF-2DA) in a solution containing 20 mM Hepes-KOH, pH 7.5. Representative images showing endogenous NO concentrations in root apical cells treated with various chemicals (control, 100 μM $AlCl_3$, 100 μM SNP, 10 mM c-PTIO, 10 mM L-NNA, and 10 mM tungstate) for 20 min detected by confocal laser scanning microscopy are given in (a). Mean DAF-2DA fluorescence densities relative to the fluorescence densities measured in control solutions determined from images shown in (a) are given in (b). Data are mean ± SE from measurements of at least four roots for each treatment. Bar, 200 μm (a). This figure is reproduced from Tian *et al.* (2008) [29].

9.2.2 NO Generated from NR Under Abiotic Stress

It is apparent that even if NOS serves as a source of NO in plants, alternative sources must also be present, with nitrate reductase considered the most likely candidate [30]. In fact, over 20 years ago, nitrogen oxides were reported in *in vivo* assays of soybean leaves [31]. Using nitrate reductase-deficient mutants of soybean, Dean and Harper [32] found that such plants did not evolve NO, unlike wild-type plants, indicating NR as a likely enzyme candidate for NO production. These workers isolated and characterized soybean NR activity, showing that it was NAD(P)H-dependent, had a pH optimum of 6.75, and was cyanide-sensitive.

NR is a key enzyme of nitrate assimilation in higher plants [29], often catalyzing the rate-limiting step. NR uses NAD(P)H as an electron source for the conversion of nitrate to nitrite [33]. NR also has the capacity to generate NO, an activity that has been demonstrated both *in vitro* [34] and *in vivo* [35]. This enzyme can generate NO from nitrite (NO^{2-}) with NADH as electron donor, and the catalysis site is probably the molybdenum cofactor (Moco) [35, 36]. Nitrate reductase also produces peroxinitrite simultaneously with NO [34]. However, the NO production capacity of nitrate reductase at saturating NADH and nitrite concentrations is about 1% of its nitrate reduction capacity, and *in vivo* NO production depends on the total nitrate reductase activity, the enzyme activation state, and the intracellular accumulation of NO^{2-} and NO^{3-} [35].

Some evidence has shown that NO generated from NR also takes part in the signal of abiotic stress. In *Arabidopsis*, NR is encoded by two genes, *NIA1* and *NIA2*. Recently, Modolo *et al.* [37] reported that a lack of NR activity in the *nia1nia2* double mutant leads to reduced levels of L-arginine and that exogenous L-arginine can restore NO generation in this mutant. This suggests that aberrant nitrogen metabolism due to a lack of NR might somehow affect NO production via an arginine-dependent process. However, in *Arabidopsis* guard cells, it appears that the NIA1 isoform of NR preferentially accounts for NO synthesis during ABA-induced stomatal closure and does so in a background where nitrogen metabolism is unlikely to be disturbed. The result of Desikan's study [38] also showed that NR-mediated NO synthesis is required for ABA-induced stomatal closure. In their experiment, the authors provided pharmacological, physiological, and genetic evidence that NO synthesis in *Arabidopsis* guard cells is mediated by nitrate reductase. Maize NR (NR, EC 1.6.6.1) is a KCN-sensitive enzyme [39]. Tungstate serves as a molybdenum analogue, and the reduction in NR activity in plants is caused by the synthesis of an inactive tungstoprotein [40]. Water stress-induced NO production was sensitive to pretreatment with tungstate. Similar results were obtained using KCN and NaN_3, two potent inhibitors of NR [36, 41]. Pretreatments with KCN and NaN_3 blocked NO production induced by water stress [17]. However, tungstate, KCN, and NaN_3 are nonspecific inhibitors in plants. KCN and NaN_3 are peroxidase (POX) inhibitors, and tungstate is also the inhibitor of ABA synthesis. These results suggest that NR might be the source of water stress-induced NO generation. The concentration-, time-, and tissue-dependent generation of NO through NR was investigated in roots of *Pisum sativum*

L. and *Triticum aestivum* L. under osmotic stress, as well as in *Petroselinum crispum* L. under drought stress [42]. The results indicated that NR is one of the sources of NO under these stress conditions. In roots, NR activity is generally low as it was in the control samples; however, it increased during osmotic treatments suggesting its role in osmotic reactions.

In addition to the above results, NaCl stress also induced NR to generate NO [43]. The results of this study showed that glucose-6-phosphate dehydrogenase (G-6-PDH) activity was enhanced rapidly in the presence of NaCl and reached a maximum at 100 mM. NO production and NR activity were also induced by 100 mM NaCl. NO production was reduced by NaN_3 (an NR inhibitor) but was not affected by L-NNA (an NOS inhibitor). Application of 2.5 mM Na_3PO_4, an inhibitor of G-6-PDH, blocked the increase of G-6-PDH and NR activity, as well as NO production in red kidney bean roots under 100 mM NaCl. Activities of antioxidant enzymes in red kidney bean roots increased in the presence of 100 mM NaCl or sodium nitroprusside (SNP), an NO donor. The increased activities of all antioxidant enzymes tested at 100 mM NaCl were completely inhibited by 2.5 mM Na_3PO_4. On the basis of these results, the authors concluded that G-6-PDH plays a pivotal role in NR-dependent NO production and in establishing tolerance of red kidney bean roots to salt stress.

9.3 NO Signaling Functions in Abiotic Stress Responses

NO has been reported to be induced rapidly by several types of chemical, mechanical, and environmental stressors in a variety of plant species, and to regulate plant responses to abiotic stresses.

It is well known that various abiotic stresses such as drought, low and high temperatures, and UV and ozone exposure induce the generation of free radicals and other oxidants, particularly from the chloroplasts, mitochondria, and peroxisomes [44], resulting in oxidative stress, that is, an increased level of reactive oxygen species in cells [45]. ROS not only initiate several oxidatively destructive processes but also trigger various signaling pathways [46, 47]. In fact, NO interacts with ROS in various ways and might serve an antioxidant function during various stresses [48]. Modulation by NO of superoxide formation [49] and inhibition of lipid peroxidation [50] also illustrate its potential antioxidant role, mainly due to its ability to maintain the cellular redox homeostasis and regulate ROS toxicity. Moreover, NO was also proposed to eliminate excess nitrite from plant cells as a high concentration of nitrite is toxic to plant cells [51, 52]. On the other hand, excess NO can result in nitrosative stress, so a favorable balance of ROS/NO is important. Another key role of NO in abiotic stress response relies on its properties as a signaling molecule. NO is involved in the signaling pathway downstream of jasmonic acid synthesis and upsteam of H_2O_2 synthesis, and regulates the expression of some genes involved in abiotic stress tolerance [53, 54]. A synergistic effect occurs between NO and ROS in ABA biosynthesis [55]. Furthermore, NO influences Ca^{2+} level in response to either salinity or osmotic stress caused by sorbitol [21, 56]. In the following sections, the means by which NO is associated with each major environmental stress is discussed in greater detail.

9.3.1
Function of NO Under Drought Stress

From the point of view of plant productivity, drought stress is especially important. Drought stress is a major environmental constraint on crop productivity and performance, and understanding the cellular processes that ameliorate drought stress and conserve water is clearly important. As reported earlier, drought promoted NO production in pea and tobacco [56, 57].

Exogenously applied SNP, an NO donor, reduced water loss from detached wheat leaves and seedlings subjected to drought conditions, decreased ion leakage and transpiration rate, and induced stomatal closure, thereby enhancing plant tolerance to drought stress [58]. Interestingly, a specific NO scavenger, carboxy-PTIO, reverted the above actions of NO. Results of this experiment suggest that exogenous application of NO donors might confer on plants an increased tolerance to severe drought stress conditions. It was shown that treatment of plants with exogenous NO enhanced drought tolerance of cut leaves and seedlings of wheat [59]. These results showed that treatment with PEG, termed as drought stress, induced accumulation of hydrogen peroxide and resulted in lipid peroxidation. On the other hand, activities of superoxide dismutase (SOD), CAT, and PAL increased under mild stress to counteract the oxidative injury and then decreased when the stress became severe. A level of 0.2 mM SNP treatment enhanced wheat seedling growth and maintained relatively high water content, and alleviated oxidative damage. However, 2 mM SNP aggravated the stress as a result of uncontrolled generation of reactive oxygen species and ineffectiveness of antioxidant systems.

In a study by Sang *et al.* [17], the sources of NO production under water stress, the role of NO in water stress-induced H_2O_2 accumulation, and subcellular activities of antioxidant enzymes in leaves of maize (*Zea mays* L.) plants were investigated. Water stress-induced increases in the production of NO were blocked by pretreatments with inhibitors of NOS and NR, suggesting that NO is produced from NOS and NR in leaves of maize plants exposed to water stress. Water stress also induced increases in activities of the chloroplastic and cytosolic antioxidant enzymes such as superoxide dismutase, ascorbate peroxidase (APX), and glutathione reductase (GR). Increases in activities of antioxidant enzymes were reduced by pretreatments with inhibitors of NOS and NR. Exogenous NO increased the activities of water stress-induced subcellular antioxidant enzymes, which decreased accumulation of H_2O_2. These results suggest that NOS and NR are involved in water stress-induced NO production and NOS is the major source of NO. The potential ability of NO to scavenge H_2O_2 is at least in part due to the induction of a subcellular antioxidant defense mechanism.

Our experimental results [19] showed that NO dependence on an NOS activity is involved in the signaling of drought-induced protective responses in maize seedlings. Both NOS activity and rate of NO release substantially increased under dehydration stress. The high NOS activity induced by c-PTIO as NO scavenger and NO accumulation inhibited by NOS inhibitor L-NAME in dehydration-treated maize seedlings indicated that most NO production due to water deficit stress may be generated from NOS-like activity. Exogenous NO (SNP) treatment alleviated water loss and oxidative damage in maize leaves under water deficit stress. When c-PTIO as a specific NO scavenger was applied, the effects of applied SNP were counteracted. Treatment with L-NAME on leaves also led to a higher membrane permeability, higher transpiration rate, and lower SOD activities than those of control leaves, indicating that NOS-like activity was involved in the antioxidative defense under water deficit stress. These results suggested that NO dependence on NOS-like activity serves as a signaling component in the induction of protective responses and is associated with drought tolerance in maize seedlings.

ABA is synthesized following turgor loss and stimulates guard cell NO synthesis, but the effects of dehydration on NO generation have not yet been resolved. There is also some evidence suggesting that ROS and NO interact to induce ABA biosynthesis. In response to drought stress, an increase in NOS-like activity was observed in wheat seedlings, and ABA accumulation was inhibited by NOS inhibitors [55]. ABA-induced NADPH oxidase activity occurring during drought stress, leading to increased ROS levels, has also been reported for maize [60, 61], indicating a close interplay between ABA, ROS, and NO levels. The study of Zhang *et al.* [18] suggest that ABA-induced H_2O_2 production mediates NO generation, which, in turn, activates mitogen-activated protein kinases (MAPKs) and results in upregulation in the expression and activities of antioxidant enzymes in ABA signaling. The use of ABA, ROS, and NO mutants should help to elucidate these complex interactions.

The protective effect of NO in osmotic stress was recently confirmed in two ecotypes of reed suspension cultures. The results of Zhao *et al.* [62] suggest that polyethylene glycol (PEG-60000)-induced NO release in stress-tolerant, but not sensitive, ecotype reed can effectively protect against oxidative damage and confers an increased tolerance to osmotic stress.

9.3.2 Function of NO Under Salt Stress

Soil salinity, similar to drought stress, is a major constraint to crop production because it limits crop yield and restricts the use of the previously uncultivated land. The involvement of NO in salt tolerance has drawn much attention in the past few years. The NO function in salt tolerance has been demonstrated in many plant species.

Pretreatment with NO donor SNP protected young rice seedlings, resulting in better plant growth and viability [63], promoted seed germination and root growth of yellow lupine seedlings [64], and increased growth and dry weight of maize seedlings [65] under salt stress conditions. The results of Ruan *et al.* [66] showed that changes in chlorophyll and malondialdehyde (MDA) contents and plasma membrane permeability confirmed that SNP could markedly alleviate oxidative damage to wheat (*T. aestivum* L.) leaves induced by NaCl treatment. Further results demonstrated that NO significantly enhanced activities of SOD and CAT, both of which separately contributed to the delay of O^{2-} and H_2O_2 accumulation in wheat leaves under salt stress. These results therefore suggest that NO could strongly protect wheat leaves from oxidative damage caused by salt stress. Another result of Ruan *et al.* [67] indicated that exogenous NO releaser SNP was involved in abscisic acid-induced proline accumulation in wheat seedling leaves under salt stress. They also confirmed that Ca^{2+} was an important intermediate in the NO signaling pathway in proline accumulation under salinity conditions. Shi *et al.* [68] also reported that exogenous NO protects cucumber roots against oxidative damage induced by salt stress.

Zhang *et al.* [69] reported that NO enhanced salt tolerance in maize seedlings, by increasing K^+ accumulation in roots, leaves, and sheathes, while decreasing Na^+ accumulation. In further studies, the same group showed that both NO and NaCl treatment stimulated vacuolar H^+-ATPase and H^+-PPase activities, resulting in increased H^+-translocation and Na^+/H^+ exchange. Furthermore, NaCl-induced H^+-ATPase and H^+-PPase activities were diminished by NO scavenger MB-1 [65]. Similarly, NO-induced salt resistance of callus from *Populus euphratica* under salt stress also by increasing the K^+/Na^+ ratio; this process was mediated by H_2O_2 and depended on increased plasma membrane H^+-ATPase activity [69]. In addition, NO was observed to stimulate the expression of plasma membrane H^+-ATPase in both salt-tolerant and salt-sensitive reed callus [21]. The results of Shi [68] also showed that application of SNP alleviated inhibition of H^+-ATPase and H^+-PPase in PM and/or tonoplast by NaCl. Application of sodium ferrocyanide (an analogue of SNP that does not release NO) did not show the effect of SNP; furthermore, the effects of SNP were reversed by addition of hemoglobin (an NO scavenger).

Arabidopsis mutant *Atnoa1* with an impaired *in vivo* NO synthase activity and a reduced endogenous NO level was more sensitive to NaCl stress than the wild type [20]. When grown under NaCl stress, the wild-type *Arabidopsis* plants exhibited a higher survival rate than *Atnoa1* plants. *Atnoa1*plants had higher levels of H_2O_2 than wild-type plants under both control and salt stress, suggesting that *Atnoa1* is more vulnerable to salt and oxidative stress than wild-type plants [16]. Treatment with exogenous NO (SNP) to *Atnoa1* alleviated the oxidative damage caused by NaCl stress; inhibition of nitric oxide accumulation in the wild type plants produced opposite effects. *Atnoa1* mutants displayed a greater Na^+/K^+ ratio in shoots than wild type when exposed to NaCl; however, SNP treatment attenuated this elevation of Na^+/K^+ ratio [20]. In Fan's laboratory, expression of a rice gene *OsNOA1* homologous to *Arabidopsis AtNOA1* was found that can re-establish diminished NO synthesis in *Atnoa1*, and induced the expression of plasma membrane Na^+/H^+ antiporter gene *AtSOS1* and H^+-ATPase gene *AtAHA2*, resulting in the restoration of *Atnoa1* in terms of Na^+/K^+ ratio and salt tolerance phenotypes [70]. This phenomenon can be mimicked by exogenous application of NO donor. Along with other studies, we suggest that NO may enhance salt tolerance in plants by increasing the expression of plasma membrane Na^+/H^+ antiporter gene and H^+-ATPase gene that are required for Na^+ homeostasis and K^+ acquisition.

NO and H_2O_2 function as signaling molecules in plants under abiotic and biotic stress. Callus from *Populus euphratica*, which shows salt tolerance, was used to study the interaction of NO and H_2O_2 in plant adaptation to salt resistance [70]. Results showed that the K/Na ratio increased in *P. euphratica* callus under salt stress. Application of glucose/glucose oxidase (G/GO, a H_2O_2 donor) and SNP revealed that both H_2O_2 and NO resulted in an increased K/Na ratio in a concentration-dependent manner. Diphenylene iodonium (DPI, an NADPH oxidase inhibitor) counteracted both H_2O_2 and NO effects by increasing the Na percentage and decreasing the K percentage and thus decreasing the K/Na ratio. N^G-monomethyl-l-Arg monoacetate (NMMA) an NO synthase inhibitor, and c-PTIO, a specific NO scavenger, reversed the NO effect but did not block the H_2O_2 effect. Exogenous H_2O_2 increased the activity of PM H^+-ATPase, but the effect could not be diminished by NMMA and PTIO. The NO-induced increase in PM H^+-ATPase can be reversed by NMMA and PTIO but not by DPI. Western blot analysis demonstrated that NO and H_2O_2 stimulated the expression of PM H^+-ATPase in *P. euphratica* callus. These results indicate that NO and H_2O_2 served as intermediate molecules in inducing salt resistance in callus from *P. euphratica* under salt stress by increasing the K/Na ratio, which depended on the increased PM H^+-ATPase activity.

9.3.3
Function of NO Under Ultraviolet Radiation

Ultraviolet radiation has the potential to disproportionately affect metabolic processes in humans, animals, plants, and microorganisms. The impacts of UV-B radiation on growth, development, and metabolism of plants have been studied widely [71–74]. Some evidence has shown that NO also imparts a protective effect on plants under ultraviolet radiation. The results of Mackerness *et al.* [24] suggest that UV-B exposure leads to the generation of NO through increased NOS activity, giving rise to the upregulation of CHS gene. Results of a study by Shi *et al.* [75] suggest that NO may effectively protect plants against UV-B radiation, most probably through increased activity of antioxidative enzymes. It needs to be mentioned that NO donor treatment of potato tubers prior to UV-B irradiation resulted in the development of almost 50% more healthy leaves compared to plants not subjected to NO treatment. The study results of Qu *et al.* [25] showed that UV-B radiation significantly induced NOS activity and promoted NO release, subsequently inhibited xyloglucan-degrading activity that led to the inhibition of pea stem elongation. Zhang *et al.* [26] found that NO was a second messenger associated with developmental growth under UV-B radiation. It was found that UV-B radiation significantly induced NOS activities and accelerated the release of apparent NO of mesocotyl and that rhizospheric treatments to exogenous NO donors may mimic the response of the mesocotyl to UV-B radiation. Such effects include inhibition of mesocotyl elongation, decrease in exo- and endoglucanase activities, and increase in protein content of the cell wall of the mesocotyl. The results of Wang *et al.* [27] indicate that NO generated from NOS-like activity appeared to act in the same direction or synergistically with ROS to induce ethylene biosynthesis in a defense response to UV-B radiation in leaves of maize seedlings. He *et al.* [76] reported that UV-B radiation induces stomatal closure by promoting NO and H_2O_2 production.

9.3.4
Function of NO Under Heat and Low Temperature

NO also participates in plant response to high- and low-temperature stress. Short-term heat stress caused an increase in NO production in alfalfa [77]. Application of NO mediates resistance to chill in tomato, wheat, and maize [78]. It is possible that this effect reflects the antioxidant properties of NO, by suppressing the high levels of ROS that accumulate following exposure to chilling or heat stress [46]. Song *et al.* [79] reported that the application of two NO donors, SNP and *S*-nitroso-*N*-acetylpenicillamine (SNAP), dramatically alleviated heat stress-induced ion leakage increase, growth suppression, and cell viability decrease in reed callus under heat stress. H_2O_2 and MDA contents decreased and activities of superoxide dismutase, catalase, ascorbate peroxidase, and peroxidase increased in callus in the presence of NO donors under heat stress. The potassium salt of c-PTIO arrested NO donor-mediated protective effects. Moreover, measurement of the rate of NO release showed that NO production significantly increased in reed callus under heat stress. These results suggest that NO can effectively protect callus from oxidative stress induced by heat stress and that NO might act as a signal in activating active oxygen scavenging enzymes under heat stress. In the study by Uchida *et al.* [62], rice seedlings pretreated with low levels of H_2O_2 or NO permitted the survival of more green leaf tissue and resulted in higher quantum yield for photosystem II than in nontreated controls under salt and heat stress. It was also shown that pretreatment induces not only active oxygen scavenging enzyme activities but also expression of transcripts for stress-related genes encoding sucrose-phosphate synthase, D-pyrroline-5-carboxylate synthase, and small heat shock protein 26. These results suggest that H_2O_2 and NO can increase both salt and heat tolerance in rice seedlings by acting as signal molecules for the response.

9.3.5
Function of NO Under Heavy Metal Stress

In relation to other abiotic stresses, it has been documented that exogenous NO reduces the destructive action of heavy metals on plants. Kopyra and Gwóźdź [63] reported that SNP stimulates seed germination and root growth of lupin (*Lupinus luteus* L. cv. Ventus). The promoting effect of NO on seed germination persisted even in the presence of heavy metals (Pb, Cd) and sodium chloride. The inhibitory effect of heavy metals on root growth was accompanied by increased activity of SOD, which in roots preincubated with SNP was significantly higher. Changes in the activity of other antioxidant enzymes, peroxidase (EC 1.11.1.7) and catalase, were also detected. Using the superoxide anion ($O_2^{\bullet -}$)-specific indicator, dihydroethidium (DHE), the authors found intense DHE-derived fluorescence in heavy metal-stressed roots, whereas in those pretreated with SNP the fluorescence was very low, comparable to the level in unstressed roots. On the basis of the above data, they concluded that the protective effect of NO in stressed lupin roots may be at least partly due to the stimulation of SOD activity and/or direct scavenging of the superoxide anion. Singh *et al.* [80] investigated whether SNP has any ameliorating action against Cd-induced oxidative damage in plant roots and thus a protective role against Cd toxicity. Supplementation of Cd with SNP significantly reduced Cd-induced lipid peroxidation, H_2O_2 content and electrolyte leakage in wheat roots. The results indicated an ROS scavenging activity of NO. Another study evaluated the protective effect of NO against Cd-induced oxidative stress in sunflower leaves [81]. Sunflower leaves were found to significantly attenuate Cd-induced oxidative damage under Cd stress. This effect was mainly attributed to the prevention of growth inhibition and chlorophyll degradation, recovery of CAT activity and GSH levels, and enhancement of ASC content and APOX activity, as components of the antioxidant machinery that allowed the plant to better cope with metal stress. These studies indicate that NO may effectively reduce the level of ROS generated during stress and, thus, limit oxidative damage in plant cells.

In addition to reducing the level of ROS generated during heavy metal stress, NO can also increase the osmotic adjustment substance. In a study by Zhang [82], Cu-induced NO generation and its relationship with proline synthesis in *Chlamydomonas reinhardtii* was investigated. Test alga accumulated a large quantity of proline after exposure to relatively low Cu concentrations. A concomitant increase in intracellular NO level was observed with increasing Cu concentrations. Data analysis revealed that the endogenous NO generated was positively associated with proline level in Cu-stressed algae. The involvement of NO in Cu-induced proline accumulation was confirmed by using SNP and an NO scavenger c-PTIO. The results indicate that Cu-responsive proline synthesis is closely related to NO generation in *C. reinhardtii*, suggesting the regulatory function of NO in proline metabolism under heavy metal stress.

In order to demonstrate the possible role of NO in response to heavy metals in the metal accumulator *Brassica juncea* and the crop plant *P. sativum*, researchers grew these plants in the presence of 100 μM cadmium, copper, or zinc [83]. They obtained different NO levels with different heavy metal loads; the most effective metals were copper and cadmium, where the NO production doubled after 1 week of treatment. In the case of copper treatment, two-phase kinetics was found, that is, a rapid NO burst in the first 6 h was followed by a slower, gradual increase. The fast appearance of NO in the presence of cupric ions suggests that this can be a novel reaction hitherto not studied in plants under heavy metal stress. After a long-term treatment, NO levels were inversely related to nitrite concentrations that originated from nitrate reductase activity, suggesting conversion of nitrite to nitric oxide by known enzymatic pathways.

9.3.6
Function of NO Under Other Abiotic Stresses

Another atmospheric pollutant that interacts with NO is ozone. Ozone treatment of *Arabidopsis* plants induced NOS activity that preceded accumulation of salicylic acid

(SA) and cell death [84]. In tobacco, NO was found to induce SA synthesis [4]. Moreover, NO treatment has been shown to increase levels of ozone-induced ethylene production and leaf injury [30].

Wounding is a common consequence of pathogen challenge of plants, during which the generation and increased accumulation of NO and H_2O_2 are frequently observed [3]. Orozco-Cardenas and Ryan [85] demonstrated that although wounding per se does not induce the generation of NO, treatment with NO donors inhibited H_2O_2 generation following wounding, as well as the expression of specific wound-induced genes. This suggests that NO produced during pathogenesis might inhibit H_2O_2 synthesis and the activation of specific wound-induced signaling pathways.

It is worth pointing out that several forms of abiotic stress (cold and heat stress, salt and drought stress) lead to enhanced polyamine (PA) biosynthesis [2]. Tun *et al.* [86] observed that PAs induce NO generation in *A. thaliana* seedlings and concluded that NO may be a link between PA-mediated stress response and other stress mediators using NO as an intermediate.

9.4
NO Signal Transduction in Plants Under Abiotic Stress

Generally speaking, the signaling function of NO may be realized via a cGMP-dependent pathway or one independent of cGMP, that is, *S*-nitrosylation/denitrosylation of protein in mammalian systems. The mechanism of cGMP-mediated action consists of the covalent bonding of NO with the heme group of guanylate cyclase, which enhances enzymatic activity and affects the generation of cyclic GMP.

On the basis of mammalian work and emerging studies on plant systems, it is probable that both the transient activation of soluble guanylyl cyclase (sGC), which generates the second messenger cGMP, and the reversible modulation of protein activity via *S*-nitrosylation-induced conformational changes [87] are major components of NO signaling. In plant cells, as in animal cells, several signaling pathways coexist for NO-mediated signals, including cyclic nucleotides, Ca^{2+} ions, and protein kinases, as well as other, so far poorly recognized elements.

9.4.1
cGMP-Dependent Signaling

The evidence that cGMP is an NO signaling intermediate has been obtained in several *systems* [88]. Both salt and osmotic stress, two conditions that would both be expected to induce ABA synthesis, induced a rapid increase in cGMP content of *Arabidopsis* seedlings [89]. Using a sensitive radioimmunoassay technique, the cGMP content of pea epidermis and *Arabidopsis* guard cell fragments has been measured in the pmol g^{-1} range, and transient increases in cGMP levels following either ABA or SNP treatment have been observed that could be prevented by coincubation of the treated tissues with PTIO [88]. Further evidence that cGMP mediates the effects of ABA in stomatal guard cells has resulted from pharmacological work using the cell-permeable cGMP analogue 8-bromo cGMP (8BrcGMP) and inhibitors of NO-sensitive sGC such as 1*H*-[1,2,4] oxadiazolo[4,3-*a*]quinoxalin-1-one (ODQ). The ABA- or NO-induced closure of pea [90] and *Arabidopsis* [88] stomata was inhibited by ODQ. This inhibition could be prevented by coincubation of the ABA-/NO-stimulated, ODQ-treated guard cells with 8BrcGMP. However, 8BrcGMP alone did not induce stomatal closure. Thus, it would appear that although an elevated level of cGMP is required for an effective ABA-induced stomatal closure, additional signaling pathways stimulated by ABA must operate in concert for such an increase to mediate its effects.

Growing evidence suggests that NO regulates the signaling cascade via cADPR and Ca^{2+} mobilization. Nicotinamide, a potential inhibitor of cADPR synthesis, blocks ABA- and NO-induced stomatal closure [90]. Other, previously mentioned downstream targets of NO are calcium ions. NO may act through cGMP and cADPR to modulate intracellular Ca^{2+}-permeable channels to elevate free cytosolic calcium levels in cells. Garcia-Mata *et al.* [91] have also shown that NO-induced intracellular Ca^{2+} release and regulation of guard cell plasma membrane K^+ and Cl^- channels are mediated by a cGMP- and cADPR-dependent pathway. Some studies supporting a potential role for NO as an endogenous regulator of Ca^{2+} mobilization in physiological contexts have been reported. It has been shown that NO contributes to $[Ca^{2+}]$ cyt increases in plant cells exposed to abiotic stresses such as high temperatures, hyperosmotic conditions, and salinity stress [56].

9.4.2
Downstream Signaling for NO Action

The above studies also raise the question of how the NO-mediated Ca^{2+} fluxes are propagated downstream into cellular responses. Clearly, this aspect has been poorly investigated. However, it is likely that protein kinases might represent an important pathway by which NO-dependent Ca^{2+} signals are decoded. Recently, a 50-kDa calcium-dependent protein kinase (CDPK), the activity of which was induced by NO through a Ca^{2+}-dependent process, was characterized in cucumber explants [92]. The 50-kDa CDPK might contribute to NO-induced adventious root formation. Likewise, in Courtois's laboratory, activation of the tobacco MAPK SIPK (salicylic acid-induced protein kinase) by NO donors [93] was found to require a transient influx of extracellular Ca^{2+} in tobacco cells [94]. ABA and NO activate MAPKs, and MAPK activation has been suggested as a convergence point for guard cell H_2O_2 and NO signaling during ABA-induced stomatal closure [88]. The results of Zhang *et al.* also suggest that ABA-induced H_2O_2 production mediates NO generation, which, in turn, activates MAPK and results in upregulation in the expression and activities of antioxidant enzymes in ABA signaling [18].

The first evidence that NO modulates the activation of a member of the plant SNF1-related protein kinase 2(SnRK2) subfamily has been reported recently [95]. Plant SnRKs are classified into three subfamilies: SnRK1, SnRK2, and SnRK3, the SnRK2 and SnRK3 subfamilies being specific to plants [96]. Members of the

SnRK2 subfamily function in abiotic stress signaling and include the tobacco 42-kDa protein kinase NtOSAK (*Nicotiana tabacum* osmotic stress-activated protein kinase) [97]. NtOSAK is rapidly activated in response to osmotic stress through phosphorylation of two serine residues (154 and 158) located within the enzyme activation loop [98]. In Lamotte's laboratory, it has been demonstrated that the NO donor DEA/NONOate induced a rapid and transient activation of NtOSAK in tobacco suspension cell cultures [95]. Furthermore, evidence reveals that NO may be a key component of the hyperosmotic stress-induced signaling cascade leading to NtOSAK activation. An attempt was also made to clarify the NO-dependent upstream pathway of NtOSAK activation. Initial data established that neither NO-mediated Ca^{2+} influx nor Ca^{2+} release from internal stores was required for NtOSAK activation [95]. Among other possibilities, NtOSAK activity might be upregulated through phosphorylation by an upstream NO-dependent protein kinase, by autophosphorylation, and/or through direct *S*-nitrosylation or nitration by NO-derived species. Preliminary experiments do not favor the last possibility, however.

The question that subsequently comes to mind is the incidence of the NO/NtOSAK pathway on cell response. Protein kinases of the SnRK2 subfamily are activated by osmolytes and some by ABA as well, highlighting a role for these enzymes in a general response to osmotic stress [94]. The SnRK2 kinases present in guard cells, AAPK (ABA-activated protein kinase) from *Vicia faba* and its *Arabidopsis* orthologue SnRK2.6/OST1/SRK2E, play an important role in ABA signaling in response to drought and regulate stomatal closure under low-humidity stress [94]. It has been shown that the other *Arabidopsis* ABA-dependent SnRK2 kinase, SRK2C/SnRK2.8, improves plant drought tolerance, probably by promoting the upregulation of stress-responsive gene expression, including DREB1A/CBF3 encoding a transcription factor that broadly regulates stress-responsive genes [99]. Several lines of evidence indicate that SnRK2 kinases also phosphorylate and, in this way, activate transcription activators AREB1 and TRAB1 in *Arabidopsis* and rice, respectively [100, 101]. These data strongly suggest that SnRK2 protein kinases are involved in the regulation of expression of ABA-responsive genes. Based on these studies, it is plausible that plant cells challenged by osmotic stress might use NO as an early signaling compound acting upstream of SnRK2-induced pathways.

The above studies support a link between NO, cADPR, cGMP, and Ca^{2+} and protein kinase. Mechanisms through NO might affect transduction processes implying the regulation of key signaling proteins such as protein kinases and Ca^{2+}-permeable channels as well as the mobilization of second messengers including Ca^{2+}, cGMP, and cADPR (Figure 9.2). However, little is known at the molecular level concerning these signaling proteins, and important goals are to identify them and to investigate how NO modulates their activities. Furthermore, it is necessary to define the physiological relevance of these modulations and to understand how interplays between NO and Ca^{2+} guide the cell toward a specific response. Such tasks will require functional analysis of the molecular mechanisms that relay NO-dependent Ca^{2+} signals under abiotic stress. Finally, pharmacological evidence for cADPR involvement in mediating NO induced Ca^{2+} mobilization has been obtained, but the direct measurement of cellular cADPR levels under abiotic stress is urgently

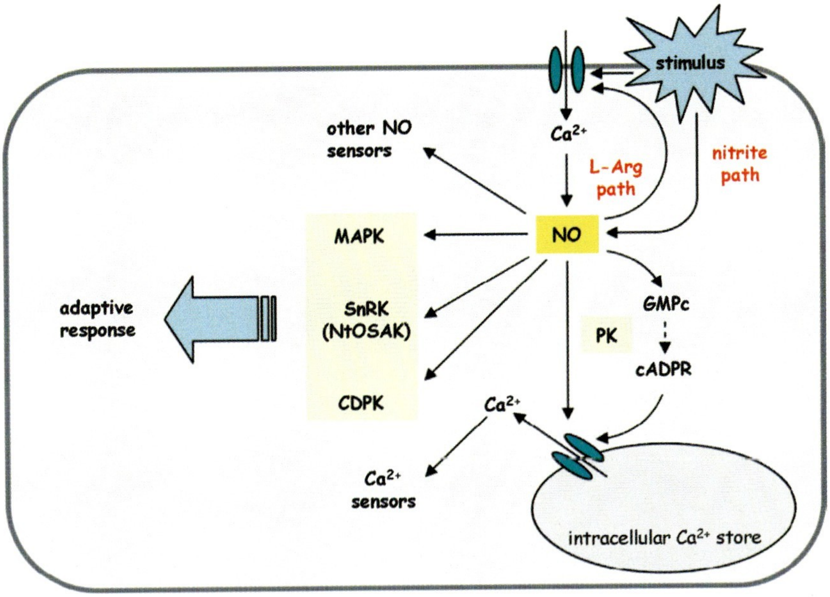

Figure 9.2 Interplays between NO, Ca^{2+}, and protein kinases in plant cells. NO production occurs through two enzymatic routes: a nitrite-dependent and an L-arginine (L-arg) route. The L-arginine-dependent pathway is upregulated by upstream Ca^{2+} fluxes that might be partly mediated by CNGCs. The increases in $[Ca^{2+}]$cyt caused by NO are due to extracellular Ca^{2+} uptake and/or mobilization of intracellular Ca^{2+}. Mechanisms through which NO mobilizes intracellular Ca^{2+} might involve cADPR, cGMP, and phosphorylation-dependent processes. The NO/Ca^{2+} information appears to be partly processed by CDPKs and MAPKs. In tobacco, NO-sensor protein kinases include NtOSAK, a member of the SnRK2 subfamily activated in response to osmotic stress. Other NO sensors correspond to nitrosylated proteins. The cross talks operating between NO, Ca^{2+}, and protein kinases in plant cells exposed to biotic and abiotic stimuli (including auxin, ABA, osmotic stress, pathogens, and elicitors of defense responses) might have important functional implications such as the dynamic regulation of gene expression. As described in animal cells, NO might also display a dual role in controlling $[Ca^{2+}]$cyt concentration since both NO-dependent activation and inhibition of extracellular Ca^{2+} uptake are reported. PK: protein kinases. This figure is reproduced from Courtois *et al.* (2008) [102].

required. Although in its infancy, research into the signaling functions of NO in plants is advancing rapidly and there should soon be a much better understanding of this most unusual signaling agent.

9.5 Interactions of NO Signaling with Other Signaling Molecules in Plant Response to Abiotic Stress

It is commonly observed that NO and ROS such as superoxide and H_2O_2 are generated in response to similar stimuli and with similar kinetics. NO and ROS interact in various ways. In several situations, such as during pathogen challenge and stomatal closure induced by the hormone abscisic acid, both H_2O_2 and NO appear to be generated and function concurrently [103]. Moreover, all three signals can induce the generation of antioxidant activity that ameliorates oxidative stress [104]. Water stress or, strictly speaking, water deficit stress, which is often referred to as drought stress, is a major abiotic condition that dramatically impacts plant and crop growth and yield. Water stress occurs in plants growing in drying soil as the water lost from the leaves exceeds that taken up by the roots and results in cellular dehydration, damage, and ultimately death. Cellular dehydration can also occur during exposure of plants to other abiotic stresses that restrict water supply, such as during cold and salt stress or during anaerobic conditions resulting from root flooding (Figure 9.3). Several defense responses are activated by water stress. One of the most important of these is stomatal closure induced by ABA redistribution and synthesis.

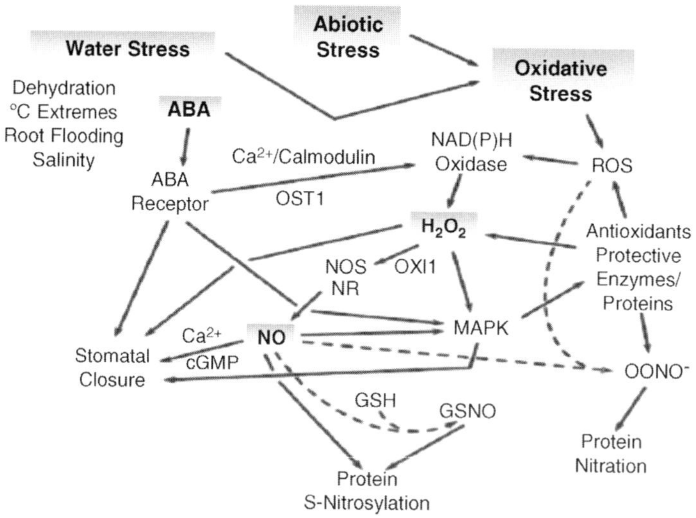

Figure 9.3 Water, abiotic, and oxidative stress and the signaling interactions between ABA, H_2O_2, and NO that occur to mediate plant survival under adverse conditions. Water stress resulting in the loss of cell turgor triggers ABA biosynthesis. ABA activates H_2O_2 generation by NAD(P)H oxidase via a signaling pathway involving the ABA receptor(s), Ca^{2+}/calmodulin, the OST1 protein kinase, and other unidentified components. H_2O_2 induces NO generation by nitrate reductase and NOS-like enzyme(s) via an as yet to be fully characterized signaling pathway that may include the OXI1 protein kinase and involve Ca^{2+}. NO induces stomatal closure through steps that require MAPKs, cGMP, and Ca^{2+}. It is also likely that NO-independent signaling from ABA and H_2O_2 also occur to cause stomatal closure during certain conditions. NO also enhances antioxidant gene and enzyme activity via MAPK and other unidentified signaling pathways. For example, superoxide dismutase activity may increase along with that of catalase and ascorbate peroxidase to combat increases in ROS, and proteins such as the dehydrins may be produced to ameliorate the effects of cell dehydration. Other abiotic stresses induce oxidative stress, the generation of H_2O_2 and NO, and enhanced antioxidant defenses. NO may also signal by inducing conformational changes in proteins as a result of *S*-nitrosylation or nitration. However, the exact role of these processes in stomatal closure and stress amelioration awaits clarification. This figure is reproduced from Neill *et al.* (2008) [88].

Zhang *et al.* [18] provide the link between ABA and H_2O_2 and NO by showing that in maize (*Zea mays*) leaves, endogenous ABA synthesized in response to dehydration induces H_2O_2 production that in turn induces NO synthesis and subsequent increases in the activities of antioxidant enzymes. Moreover, the effects on antioxidant enzyme gene expression and activity require the activation of an MAPK signaling enzyme.

ABA synthesis and action are essential for plant survival during water stress. ABA is an endogenous antitranspirant that induces stomatal closure and is an activator of various processes that enhance survival in cellular dehydration [105]. ABA signaling in guard cells is especially complex, with H_2O_2, NO, and MAPKs all playing roles [104]. ABA-induced NO production in guard cells depends on H_2O_2 generation [106]. In their previous work, Zhang *et al.* [107] demonstrated that in maize leaves water stress-induced ABA activates H_2O_2 generation via the activation of an NADPH oxidase-like enzyme, the same enzyme that generates H_2O_2 in response to ABA in *Arabidopsis* guard cells. In another study, NO was found to be an essential intermediate in ABA-regulated processes during water stress in leaf mesophyll cells as well as in guard cells. Zhang *et al.* [18] used the fluorescent dye DAF-2DA to show that both ABA and H_2O_2 induce NO generation in maize mesophyll cells. The NO scavenger cPTIO and the non-NO reactive 4AF-DA were used to demonstrate that fluorescence increases were indeed attributable to NO. ABA- and H_2O_2-induced increases in fluorescence were rapid and dose-dependent and induction by ABA was prevented by DPI, a known inhibitor of NADPH oxidase. Rapid removal of any ABA-induced H_2O_2 by pretreatment with H_2O_2 scavengers also prevented NO increase, demonstrating that H_2O_2 generation and action are required for NO production. Furthermore, osmotic stress (induced by incubation in polyethylene glycol) similarly induced NO generation that was also prevented by pretreatment with H_2O_2 scavengers. NO generation was not induced in the ABA-deficient *vp5* mutant by osmotic stress but could be activated by ABA – thereby confirming that endogenous ABA is responsible and required for H_2O_2-mediated NO production.

ABA also modulates the expression of gene networks that control other ameliorative responses. These include the maintenance of root water uptake, synthesis of osmoprotective proteins such as dehydrins, and various metabolic changes (Figure 9.3) [88]. Oxidative stress is a common feature of several abiotic stresses including water stress [108]. During oxidative stress, the redox balance of cells is disturbed by increases in the rate of generation of reactive oxygen species such as the superoxide anion (O^{2-}) and hydrogen peroxide (H_2O_2) above that of their removal by antioxidant enzymes or by reaction with antioxidant molecules. Cell functions are altered during oxidative stress not only because of oxidative damage per se but also because ROS themselves are centrally important signaling molecules. Thus, excessive quantities of ROS result in aberrant cell signaling [108]. The activation of cellular antioxidant systems is a common feature of oxidative stress, and there is increasing evidence indicating that NO is a critical factor in such responses.

Zhang *et al.* [107] had shown that osmotic stress, ABA, and H_2O_2 enhance the expression of several antioxidant genes such as *CAT1*, cytosolic ascorbate peroxidase (*cAPX*) and plastidial glutathione reductase 1(*GR1*), and the total enzyme activities of CAT, APX, GR, and SOD. In another study [18], they demonstrated that NO is an essential intermediate in these ABA and H_2O_2 enhancements. Pretreatment with the NO scavenger c-PTIO substantially (but not completely) prevented increases in gene expression and enzyme activity. Moreover, treatment with the NO donor sodium nitroprusside essentially reproduced the effects of ABA or H_2O_2. Importantly, the removal of the NO released from SNP with c-PTIO prevented the increases, and treatment with sodium ferricyanide (a molecule similar to SNP but which does not release NO) had no effect. A number of studies have already shown that exogenously applied NO can impart protective antioxidant properties. Some recent work has indicated that endogenous NO induces antioxidant defenses, potentially via ABA signaling [79, 109].

The results of Zhang *et al.* [18] also showed that MAPK activation is similarly targeted by H_2O_2 and NO in mesophyll cells and that this MAPK activation is required for downstream signaling to enhance antioxidant gene expression and enzyme activity. Both ABA and H_2O_2 activate an MAPK enzyme in maize leaves (or at least an enzyme with properties characteristic of MAPKs), but this activation is largely prevented by removal of NO with the NO scavenger cPTIO. Moreover, as with enhancement of antioxidant activity, the MAPK is activated by treatment with NO (supplied via SNP). Finally, inhibition of MAPK activation by treatment with PD98059 (an inhibitor of mitogen-activated protein kinase kinase, MAPKK) and U0126, compounds that inhibit MAPK kinases and upstream activators of MAPKs, inhibits increase in antioxidant gene expression and activity. She and Song's [110] study indicated that both 20-amino-30-methoxyflavone (PD98059) and trifluoperazine (TFP) (a specific inhibitor of CDPK) reduced levels of NO in guard cells and significantly reversed darkness-induced stomatal closure, implying that MAPKK/CDPK mediates darkness-induced stomatal closure by enhancing NO levels in guard cells. In addition, as with NO scavenger c-PTIO, but not with L-NAME, PD98059 and TFP reduced NO contents in guard cells not only induced by SNP in light but also generated during a dark period. They also reversed stomatal closure by SNP and by darkness, suggesting MAPKK and CDPK are probably related to restraining the NO scavenging to elevate NO levels in guard cells during darkness-induced stomatal closure. The results also showed that both PD98059 and TFP reduced stomatal closure by SNP, implying that the possibility of MAPKK and CDPK acting as the target downstream of NO should not be ruled out. There may be a causal and interdependent relationship between MAPKK/CDPK and NO in darkness-induced stomatal closure, and in the process this cross talk may lead to the formation of a self-amplification loop about them.

The data presented indicate a key "ABA–H_2O_2–NO–MAPK–antioxidant survival cycle" and suggest that during water stress ABA has several ameliorative functions that involve NO as a key signaling intermediate and which include the rapid induction of stomatal closure to reduce transpirational water loss and the activation of antioxidant defenses to combat oxidative damage (Figure 9.3).

These experiments indicate that endogenous NO is a key factor in the tolerance of cells to oxidative stress induced by a range of abiotic conditions including water stress, and this probably involves the enhanced expression of genes encoding antioxidant enzymes. They suggest that during water stress ABA has several ameliorative functions that involve NO as a key signaling intermediate and include the rapid induction of stomatal closure and activation of antioxidant defenses.

Of course, there are numerous unanswered questions and important areas for further research. The mechanisms by which NO is generated are still largely unresolved. Elucidation of how NO is generated by different plant cells, in different situations, is clearly a research priority.

Another question to be resolved is related to the mechanism(s) by which NO (and for that matter H_2O_2) is perceived by cells. Transient elevation of the second messenger cyclic GMP via activation of a guanylyl cyclase enzyme is a possibility, as is direct activation by reversible *S*-nitrosylation of cysteine residues, perhaps located in enzymes, transcription factors, or ion channels [89]. Functions of many other components of the ABA–H_2O_2–NO–cellular response signal transduction chain also require clarification. These include the protein kinases OST1 [111] and OXI1 [112] and many other as-yet-unknown proteins and signaling molecules. There is much work to do and much to discover in this area of plant biology.

References

1 García-Mata, C. and Lamattina, L. (2002) *Plant Physiol.*, **128**, 790–792.

2 Arasimowicz, M. and Floryszak-Wieczorek, J. (2007) *Plant Sci.*, **172**, 876–887.

3 Delledonne, M., Xia, Y., Dixon, R.A., and Lamb, C. (1998) *Nature*, **394**, 585–588.

4 Durner, J., Wendehemme, D., and Klessig, D.F. (1998) *Proc. Natl. Acad. Sci. USA*, **95**, 10328–10333.

5 Courtois, C., Besson, A., Dahan, J., Bourque, S., Dobrowolska, G., Pugin, A., and Wendehenne, D. (2008) *J. Exp. Bot.*, **59**, 155–163.

6 Moncada, S., Palmer, R.M.J., and Higgs, E.A. (1991) *Pharmacol. Rev.*, **43**, 109–142.

7 Ignarro, L.J. (2000) *Nitric Oxide, Biology and Pathobiology*, Academic Press.

8 Knowles, R.G. and Moncada, S. (1994) *Biochem. J.*, **298**, 249–258.

9 Alderton, W.K., Cooper, C.E., and Knowles, R.G. (2001) *Biochem. J.*, **357**, 593–615.

10 Corpas, F.J., Barroso, J.B., Carreras, A., Valderrama, R., Palma, J.M., León, A.M., Sandalio, L.M., and del Rio, L.A. (2006) *Planta*, **224**, 246–254.

11 Guo, F.Q., Okamoto, M., and Crawford, N.M. (2003) *Science*, **302**, 100–103.

12 Crawford, N.M., Galli, M., Tischner, R., Heimer, Y.M., Okamoto, M., and Mack, A. (2006) *Trend Plant Sci.*, **11**, 526–527.

13 Zemojtel, T., Frohlich, A., Palmieri, M.C., Kolanczyk, M., Mikula, I., Wyrwicz, L.S., Wanker, E.E., Mundlos, S., Vingron, M., Martasek, P., and Durner, J. (2006) *Trends Plant Sci.*, **11**, 524–525.

14 He, Y.K., Tang, R.H., Hao, Y., Stevens, R.D., Cook, C.W., Ahn, S.M., Jing, L.F., Yang, Z.G., Chen, L.G., Guo, F.Q., Fiorani, F., Jackson, R.B., Crawford, N., and Pei, Z.M. (2004) *Science*, **305** (5692), 1968–1971.

15 Zeidler, D., Zahringer, U., Gerber, I., Dubery, I., Hartung, T., Bors, W., Hutzler, P., and Durner, J. (2004) *Proc. Natl. Acad. Sci. USA*, **101**, 15811–15816.

16 Zhao, M.G., Zhao, X., Wu, Y.X., and Zhang, L.X. (2007) *J. Plant Physiol.*, **164**, 737–745.

17 Sang, J.R., Jiang, M.Y., Lin, F., Xu, S.C., and Zhang, A.Y., Tan, M.P. (2008) *J. Integr. Plant Biol.*, **50**, 231–243.

18 Zhang, A.Y., Jiang, M.Y., Zhang, J.H., Ding, H.D., Xu, S.C., Hu, X.L., and Tan, M.P. (2007) *New Phytol.*, **175**, 36–50.

19 Hao, G.P., Xing, Y., and Zhang, J.H. (2008) *J. Integr. Plant Biol.*, **50**, 435–442.

20 Zhao, M.G., Tian, Q.Y., and Zhang, W.H. (2007) *Plant Physiol.*, **144**, 206–217.

21 Zhao, L.Q., Zhang, F., Guo, J.K., Yang, Y.L., Li, B.B., and Zhang, L.X. (2004) *Plant Physiol.*, **134**, 849–857.

22 Trapido-Rosenthal, H.G., Zielke, S., Owen, R., Buston, L., Boein, B., Bhagooli, R., and Archer, J. (2005) *Biol. Bull.*, **208**, 3–6.

23 Bouchard, J.N. and Yamasaki, H. (2008) *Plant Cell Physiol.*, **49**, 641–652.

24 Mackerness, S.A.-H., John, C.F., Jordan, B., and Thomas, B. (2001) *FEBS Lett.*, **489**, 237–242.

25 Qu, Y., Feng, H.Y., Wang, Y.B., Zhang, M.X., Cheng, J.Q., Wang, X.L., and An, L.Z. (2006) *Plant Sci.*, **170**, 994–1000.

26 Zhang, M.X., An, L.Z., Feng, H.Y., Chen, T., Chen, K., Liu, Y.H., Tang, H.G., Chang, J.F., and Wang, X.L. (2003) *Photochem. Photobiol.*, **77**, 219–225.

27 Wang, Y.B., Feng, H.Y., Qu, Y., Cheng, J.Q., Zhao, Z.G., Zhang, M.X., Wang, X.L., and An, L.Z. (2006) *Environ. Exp. Bot.*, **57**, 51–61.

28 Garcês, H., Durzan, D., and Pedroso, M.C. (2001) *Ann. Bot. (Lond.)*, **87**, 567–574.

29 Tian, Q.Y., Sun, D.H., Zhao, M.G., and Zhang, W.H. (2007) *New Phytol.*, **174**, 322–331.

30 Neill, S.J., Desikan, R., and Hancock, J.T. (2003) *New Phytol.*, **159**, 11–35.

31 Harper, J.E. (1981) *Plant Physiol.*, **68**, 1488–1493.

32 Dean, J.V. and Harper, J.E. (1986) *Plant Physiol.*, **82**, 718–723.

33 Lea, P.J. (1999) Nitrate assimilation, in *Plant Biochemistry and Molecular Biology* (eds P.J. Lea and R.C. Leegood), John Wiley & Sons Ltd, London, UK, pp. 163–192.

34 Yamasaki, H. and Sakihama, Y. (2000) *FEBS Lett.*, **468**, 89–92.

35 Rockel, P., Strube, F., Rockel, A., Wildt, J., and Kaiser, W.M. (2002) *J. Exp. Bot.*, **53**, 103–110.

36 Yamasaki, H. (2000) *Philos. Trans. R. Soc. Lond.*, **355**, 1477–1488.

37 Modolo, L.V., Augusto, O., Almeida, I.M.G., Pinto-Maglio, C.A.F., Oliveira, H.C., Seligman, K., and Salgado, I. (2006) *Plant Sci.*, **171**, 34–40.

38 Desikan, R., Griffiths, R., Hancock, J., and Neill, S. (2002) *Proc. Natl. Acad. Sci. USA*, **99**, 16314–16318.

39 Solomonson, L.P. and Barber, J.M. (1989) *Molecular and Genetic Aspects of Nitrate Assimilation* (eds J.L. Wary and J.R. Kinghorn), Oxford Science Publications, Oxford, pp. 88–100.

40 Notton, B.A. and Hewitt, E.J. (1971) *Biochem. Biophys. Res. Commun.*, **44**, 702–710.

41 Sakihama, Y., Nakamura, S., and Yamasaki, H. (2002) *Plant Cell Physiol.*, **43**, 290–297.

42 Kolbert, Z., Bartha, B., and Erdei, L. (2005) *Acta Biol. Szeg.*, **49** (1–2), 13–16.

43 Liu, Y.G., Wu, R.R., Wan, Q., Xie, G.J., and Bi, Y.R. (2007) *Plant Cell Physiol.*, **48** (3), 511–522.

44 Mano, J. (2002) *Oxidative Stress in Plants* (eds D. Inze and M. Van Montagu), Taylor and Francis, London, pp. 217–246.

45 Mittler, R. (2002) *Trends Plant Sci.*, **7**, 405–410.

46 Neill, S.J., Desikan, R., Clarke, A., Hurst, R.D., and Hancock, J.T. (2002) *J. Exp. Bot.*, **53**, 1237–1242.

47 Vranova, E., Inze, D., and Van Breusegem, F. (2002) *J. Exp. Bot.*, **53**, 1227–1236.

48 Beligni, M.V. and Lamattina, L. (1999) *Trends Plant Sci.*, **4**, 299.

49 Caro, A. and Puntarulo, S. (1998) *Physiol. Plant.*, **104**, 357–364.

50 Boveris, A.D., Galatro, A., and Puntarulo, S. (2000) *Biol. Res.*, **33**, 159–165.

51 Wellburn, A.R. (1990) *New Phytol.*, **115**, 395–429.

52 Shingles, R., Roh, M.H., and McCarty, R.E. (1996) *Plant Physiol.*, **112**, 1375–1381.

53 Wendehenne, D., Durner, J., and Klessig, D.F. (2004) *Curr. Opin. Plant Biol.*, **7**, 449–455.

54 Orozco-Cardenas, M.L. and Ryan, C.A. (2002) *Plant Physiol.*, **130**, 487–493.

55 Zhao, Z., Chen, G., and Zhang, C. (2001) *Aust. J. Plant Physiol.*, **28**, 1055–1061.

56 Gould, K.S., Lamotte, O., Klinguer, A., Pugin, A., and Wendehenne, D. (2003) *Plant Cell Environ.*, **26**, 1851–1862.

57 Leshem, Y.Y. and Haramaty, E. (1996) *J. Plant Physiol.*, **148**, 258–263.

58 Garcia-Mata, C. and Lamattina, L. (2001) *Plant Physiol.*, **126**, 1196–1204.

59 Tian, X. and Lei, Y. (2006) *Biol. Plant*, **50**, 775–778.

60 Lu, S.Y., Su, W., Li, H.H., and Gu, Z.F. (2009) *Plant Physiol. Biochem.*, **47**, 132–138.

61 Jiang, M.Y. and Zhang, J.H. (2002) *J. Exp. Bot.*, **53**, 2401–2410.

62 Zhao, L., He, J., Wang, X., and Zhang, L. (2008) *J. Plant Physiol.*, **165**, 182–191.

63 Uchida, A., Jagendorf, A.T., Hibino, T., and Takabe, T. (2002) *Plant Sci.*, **163**, 515–523.

64 Kopyra, M. and Gwóźdź, E.A. (2003) *Plant Physiol. Biochem.*, **41**, 1011–1017.

65 Zhang, Y.Y., Wang, L.L., Liu, Y.L., Zhang, Q., Wei, Q.P., and Zhang, W.H. (2006) *Planta*, **224**, 545–555.

66 Ruan, H.H., Shen, W.B., Ye, M.B., and Xu, L.L. (2002) *Chin. Sci. Bull.*, **47**, 677–681.

67 Ruan, H.H., Shen, W.B., and Xu, L.L. (2004) *Acta Bot. Sin.*, **46** (11), 1307–1315.

68 Shi, Q., Ding, F., Wang, X., and Wei, M. (2007) *Plant Physiol. Biochem.*, **45** (8), 542–550.

69 Zhang, Y.Y., Liu, J., and Liu, Y.L. (2004) *J. Plant Physiol. Mol. Biol.*, **30**, 455–459.

70 Qiao, W.H. and Fan, L.M. (2008) *J. Integr. Plant Biol.*, **50**, 1238–1247.

71 Zhang, F., Wang, Y., Yang, Y., Wu, H., Di, W., and Liu, J. (2007) *Plant Cell Environ.*, **30**, 775–785.

72 Strid, A., Chow, W.S., and Anderson, J.M. (1994) *Photosynth. Res.*, **39**, 475–489.

73 Santos, I., Fidalgo, F., Almeida, J.M., and Salema, R. (2004) *Plant Sci.*, **167**, 925–935. Teramura, A.H. and Sullivan, J.H. (1994) *Photosynth. Res.*, **39**, 463–473.

74 Zu, Y.Q., Li, Y., Chen, J.J., and Chen, H.Y. (2004) *J. Photochem. Photobiol. B*, **74**, 95–100.

75 Shi, S.Y., Wang, G., Wang, Y.D., Zhang, L.G., and Zhang, L.X. (2005) *Nitric Oxide*, **13**, 1–9.

76 He, J.M., Xu, H., She, X.P., Song, X.G., and Zhao, W.M. (2005) *Funct. Plant Biol.*, **32**, 237–247.

77 Leshem, Y. (2001) *Nitric Oxide in Plants*, Kluwer Academic Publishers, Dordrecht, The Netherlands.

78 Lamattina, L., Beligni, M.V., Garcia-Mata, C., and Laxalt, A.M. (2001) US Patent 6242384 B1

79 Song, L.L., Ding, W., Zhao, M.G., Sun, B.T., and Zhang, L.X. (2006) *Plant Sci.*, **171**, 449–458.

80 Singh, H.P., Batish, D.R., Kaur, G., Arora, K., and Kohli, R.K. (2008) *Environ. Exp. Bot.*, **63**, 158–167.

81 Laspina, N.V., Groppa, M.D., Tomaro, M.L., and Benavides, M.P. (2005) *Plant Sci.*, **169**, 323–330.

82 Zhang, L.P., Mehta, S.K., Liu, Z.P., and Yang, Z.M. (2008) *Plant Cell Physiol.*, **49** (3), 411–419.

83 Bartha, B., Kolbert, Z., and Erdei, L. (2005) *Acta Biol. Szeg.*, **49** (1–2), 9–12.

84 Rao, M.V. and Davis, K.R. (2001) *Planta*, **213**, 682–690.

85 Orozco-Cardenas, M.L. and Ryan, C.A. (2002) *Plant Physiol.*, **130**, 487–493.

86 Tun, N.N., Santa-Catarina, C., and Begum, T. (2006) *Plant Cell Physiol.*, **47**, 346–354.

87 Wang, Y., Yun, B.W., Kwon, E.J., Hong, J.K., Yoon, J.Y., and Loake, G.J. (2006) *J. Exp. Bot.*, **57**, 1777–1784.

88 Neill, S., Barros, R., Bright, J., and Desikan, R. (2008) *J. Exp. Bot.*, **59**, 165–176.

89 Donaldson, L., Ludidi, N., Knight, M.R., Gehring, C., and Denby, K. (2004) *FEBS Lett.*, **569**, 317–320.

90 Neill, S.J. *et al.* (2002) *Plant Phsysiol.*, **128**, 13–16.

91 Garcia-Mata, C., Gay, R., Sokolovski, S., Hills, A., Lamattina, L., and Blatt, M.R.

(2003) *Proc. Natl. Acad. Sci. USA*, **100**, 11116–11121.

92 Lanteri, M., Pagnussat, G.C., and Lamattina, L. (2006) *J. Exp. Bot.*, **57**, 1341–1351.

93 Klessig, D.F., Durner, J., and Noad, R., *et al.* (2000) *Proc. Natl. Acad. Sci. USA*, **97**, 8849–8855.

94 Courtois, C., Besson, A., Dahan, J. *et al.* (2008) *J. Exp. Bot.*, **59**, 155–163.

95 Lamotte, O., Courtois, C., Dobrowolska, G., Besson, A., Pugin, A., and Wendehenne, D. (2006) *Free Radic. Biol. Med.*, **40**, 1369–1376.

96 Harmon, A.C. (2003) *Gravitat. Space Biol. Bull.*, **16**, 83–90.

97 Mikolajczyk, M., Awotunde, O.S., Muszynska, G., Klessig, D.F., and Dobrowolska, G. (2000) *Plant Cell*, **12**, 165–178.

98 Burza, A.M., Pekala, I., Sikora, J., Siedlecki, P., Malagocki, P., Bucholc, M., Koper, L., Zielenkiewicz, P., Dadlez, M., and Dobrowolska, G. (2006) *J. Biol. Chem.*, **281**, 34299–34311.

99 Umezawa, T., Yoshida, R., Maruyama, K., Yamaguchi-Shinozaki, K., and Shinozaki, K. (2004) *Proc. Natl. Acad. Sci. USA*, **101**, 17306–17311.

100 Furihata, T., Maruyama, K., Fujita, Y., Umezawa, T., Yoshida, R., Shinozaki, K., and Yamaguchi-Shinozaki, K. (2006) *Proc. Natl. Acad. Sci. USA*, **103**, 1988–1993.

101 Kobayashi, Y., Murata, M., Minami, H., Yamamoto, S., Kagaya, Y., Hobo, T., Yamamoto, A., and Hattori, T. (2005) *Plant J.*, **44**, 939–949.

102 Courtois, C., Besson, A., Dahan, J., Bourque, S., Dobrowolska, G., Pugin, A., Wendehenne, D. (2008) *J. Exp. Bot.*, **59**, 155–163.

103 Desikan, R., Cheung, M.K., Bright, J., Henson, D., Hancock, J.T., and Neill, S.J. (2004) *J. Exp. Bot.*, **55**, 205–212.

104 Neill, S. (2007) *New Phytol.*, **175**, 4–6.

105 Zhu, J.K. (2002) *Annu. Rev. Plant Biol.*, **53**, 247–273.

106 Bright, J., Desikan, R., Hancock, J.T., Weir, I.S., and Neill, S.J. (2006) *Plant J.*, **45**, 113–122.

107 Zhang, A., Jiang, M., Zhang, J., Tan, M., and Hu, X. (2006) *Plant Physiol.*, **141**, 475–487.

108 Bailey-Serres, J. and Mittler, R. (2006) *Plant Physiol.*, **141**, 311–321.

109 Zhou, B., Guo, Z., Xing, J., and Huang, B. (2005) *J. Exp. Bot.*, **56**, 3223–3228.

110 She, X.P. and Song, X.G. (2008) *Aust. J. Bot.*, **56**, 347–357.

111 Mustilli, A.-C., Merlot, S., Vavasseur, A., Fenzi, F., and Giraudat, J. (2002) *Plant Cell*, **14**, 308–3099.

112 Rentel, M.C., Lecourieux, D., Ouaked, F., Usher, S.L., Petersen, L., Okamoto, H., Knight, H., Peck, S.C., Grierson, C.S., Hirt, H., and Knight, M.R. (2004) *Nature*, **427**, 858–861.

10
Interplay Between Nitric Oxide and Other Signals Involved in Plant Resistance to Pathogens

Jolanta Floryszak-Wieczorek and Magdalena Arasimowicz-Jelonek

Summary

One response observed in plants challenged by an invader is the induction of a defense strategy. An efficient plant response to the invading pathogen strongly depends on the ability of the host to rapidly generate signaling molecules carrying information, for example, nitric oxide, to initiate a suitable resistance mechanism. In this chapter, we examine the pattern of NO burst as the effect of pathogen recognition in *R*-specific and in nonspecific resistance related to H_2O_2 overproduction. The cooperation of NO with H_2O_2 and other signals in triggering programmed cell death is analyzed. Moreover, evidence is provided for the involvement of NO-mediated signal cascades and posttranslational protein modifications leading to the expression of defense genes. Novel data are presented concerning stress-related cross talk of NO with salicylic acid-dependent responses and jasmonic acid and/or ethylene-dependent responses to biotrophic and necrotrophic pathogens. Recently, NO has been suggested to be involved in the modulation of signaling, leading to systemic acquired resistance. Involvement of NO in the production of putative short- and long-distance compounds, as has been shown to be effective in the local and systemic protection of plants, is also discussed. Finally, we will investigate the role of NO in the recovery from disease related to stimulation of wound-healing processes.

10.1
Introduction

Nitric oxide (NO) generation is one of the earliest defense responses initiated by a plant after it recognizes a pathogen. The time required for the pathogenic microorganism to penetrate the host cell may vary considerably due to the diverse structure and morphology of organs in different plant species and the complex biology of the pathogen. Thus, the moment of recognition of the elicitor by the plant

Nitric Oxide in Plant Physiology. Edited by S. Hayat, M. Mori, J. Pichtel, and A. Ahmad

ISBN: 978-3-527-32519-1

membrane receptor is crucial since only then is the defense strategy of the plant, including NO overproduction, activated. This fact has sometimes been overlooked in studies published to date, resulting in a considerable variation of reported time of NO synthesis after inoculation in different pathogen–plant systems. Another problem is related to the selection of an appropriate measurement method for this signal, activated *in planta* at concentrations within the range of nmol g^{-1} h^{-1}. So far, most of the applied methods of NO measurement are not sufficiently accurate to determine such a low level of NO production.

10.2 NO Burst

Pharmacological, biochemical, and genetic approaches provide evidence that early overproduction of NO after challenge by a pathogen or elicitor occurs in many plant–pathogen systems. Rapid accumulation of NO in response to avirulent bacteria has been found in soybean/*Pseudomonas syringae* pv. *glicinea* [1] and *Arabidopsis* cell suspensions/*P. syringae* pv. *maculicola* [2] by hemoglobin assay. Similarly, enhanced synthesis of NO has been observed in *Arabidopsis* leaves/*P. syringae* pv. tomato carrying *avrB* or *avrRpt2* avirulence genes [3] and in oat leaves/*Puccinia coronata* pv. *avenae* avirulence strain [4] detected by DAF. Overproduction of NO was observed using DAF fluorescence techniques as well as in tobacco leaves treated with the fungal elicitor cryptogein [5, 6] and in potato tuber disks treated with hyphal wall components extracted from *Phytophthora infestans* [7].

Owing to more precise measurements of NO, it was possible to analyze the time-dependent kinetics of NO generation. Biphasic production of NO was demonstrated in the incompatible interaction between tobacco cells and *P. syringae* pv. tomato, using noninvasive online detection of NO by a combination of membrane inlet mass spectrometry (MIMS) and restriction capillary inlet mass spectrometry (RIMS) [8]. The first detectable NO burst was recorded after approximately 1 h of treatment and was followed by a much more pronounced, second burst of NO at 4–8 h. The second NO peak was probably a key element in determining avirulence in this plant–pathogen system since in the compatible interaction between soybean cells and *P. syringae* pv. *glycinea* no further rise in NO production was observed. Next, NO levels generated in wheat leaves with *Puccinia striiformis* Westend were measured using electron spin resonance (ESR) by Guo *et al.* [9]. Two NO peaks were noted in the course of the immune response, whereas in the susceptible one, NO accumulation was not observed during the early infection stage. These authors suggested that the early changes in kinetics and relative accumulation of endogenous NO were the key factors associated with host resistance. In contrast, Clarke *et al.* [2] failed to observe a biphasic rise in NO in *Arabidopsis* cultures challenged with an avirulent strain of *P. syringae* pv. *maculicola*, whereas Mur *et al.* [10] reported the same finding in tobacco leaves challenged with avirulent and virulent *P. syringae* pathovars. The latter authors, using laser photoacoustic detection (LPAD) for NO generation, found that inoculation with an avirulent strain of *P. syringae* pv. *phaseolicola* resulted in a rapid increase in NO production after approximately 40 min,

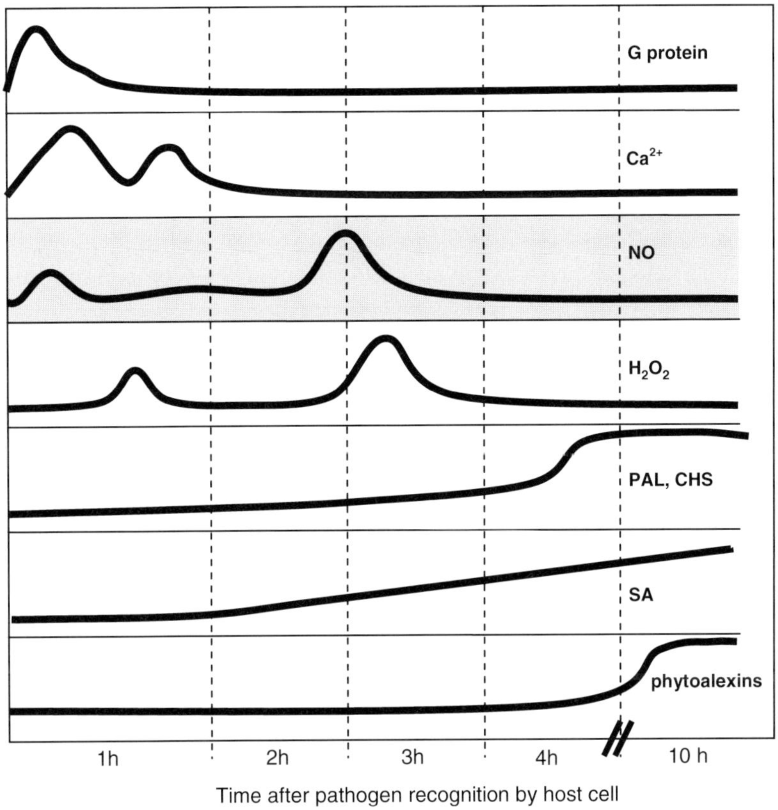

Figure 10.1 Time-dependent early events engaged in defense response after *avr–R* interaction or the application of pathogen-derived elicitors.

which lasted for 1.5 h, then temporarily decreased but still remained more pronounced than that in tobacco challenged with a virulent pathogen. In summary, the above authors concluded that NO could act as an *avr*/*R*-dependent input, modulating persistent calcium fluxes and oxidative burst, and finally influence the sequence of defense response, termed resistance scenarios [10–12].

Figure 10.1 presents time-dependent early events involved in defense responses after *avr–R* interaction or application of pathogen-derived elicitors.

Most evidence showing that NO functions as a messenger in gene-for-gene defense responses was obtained when analyzing different plant biotrophic pathogen systems [13]. It still remains to be determined what role is played by NO in the cross talk between the plant and the necrotrophic pathogen. van Baarlen *et al.* reported [14] that the generation of endogenous NO was recorded only during a compatible interaction of *Botrytis elliptica* on lily. In contrast, the participation of NO in the stimulation of rapid defense events in *Pelargonium peltatum* leaves inoculated with *Botrytis cinerea* was found by our team [15]. An almost immediate NO signal was shown in our study using real-time electrochemical NO detection with an NO-selective needle-type electrode, but the specificity of its intensity depended on the genetic makeup of the host plant; so, NO synthesis was three times higher in the resistant genotype.

On the basis of above reports, the conflicting data on NO production during recognition of an invading pathogen by a plant should be apparent. Thus, a question arises as to whether a common pattern of NO generation occurs. Probably, both the kinetics and the intensity of NO burst, even when analyzed in the identical pathogen/plant group, would change under the influence of endogenous factors of the plant (organ, vigor, and phenophase) or of the pathogen (virulence and pathogenicity), as well as different environmental factors such as temperature, light, or accompanying stress. When searching for a common denominator for the above-mentioned examples of NO generation under biotic stress, we suggest that irrespective of the type of NO generation, that is, monophasic or biphasic, it is typically stronger and more prompt in the resistant rather than the susceptible response. Since physiological effects of NO are related to concentration, concentration-dependent effects of NO can modify the sequence of defense events, for example, hypersensitive response (HR), PR gene expression, phytoalexin production, or systemic acquired resistance (SAR) establishment.

10.3 Cooperation of NO with H_2O_2 in Triggering Programmed Cell Death

The challenge of a pathogen very often leads to the induction of hypersensitive response. Until recently, it was believed that HR is found only in the case of incompatible interactions, that is, when the plant possesses a resistance gene *R* and the biotrophic pathogen a virulence gene *avr* [16]. However, it has recently been shown that HR of the host cell may also occur in the plant–necrotroph system [17] or in the case of nonhost-type resistance [18].

Although HR as one of the most striking events of plant disease resistance is well documented [19], molecular mechanisms of HR development are not yet understood in detail. It is known that cell death during HR is preceded by an oxidative burst; however, reactive oxygen species (ROS) alone are not always sufficient and additional factors, such as NO, are required to cause cell death [1, 20]. An increasing body of evidence exists indicating that NO plays a key role during plant cell death. For example, Prats *et al.* [21], using the DAF-2DA dye, observed a significant, transitory increase in NO level preceding cell death of barley epidermal cells inoculated with *Blumeria graminis* f. sp. hordei. Moreover, Zeier *et al.* [12] confirmed that NO is required in HR stimulation by obtaining a transgenic line of *Arabidopsis* with an overexpression of nitric oxide dioxygenase (NOD), an enzyme catalyzing oxidation of NO to nitrates. The transgenic *Arabidopsis* treated with an avirulent strain of *P. syringae* pv. tomato avrB showed reduced NO production and experienced a significantly inhibited cell death rate. In turn, inoculation of NR-deficient double mutants (*nia1 nia2*) of *Arabidopsis*, which lacks endogenous substrates for NO synthesis (L-arginine and NO_2^-), with an avirulent strain of *P. syringae* did induce neither NO production nor HR [22].

Hypersensitive cell death via NO is a typical example of programmed cell death (PCD) with apoptosis-like features. It has been shown that NO-donor treatment of plant tissue or cell suspension initiates chromatin condensation and DNA fragmentation as well as a loss of mitochondrial membrane electrical potential [2, 23, 24]. Moreover, NO-provoked cell death may be inhibited by the animal caspase-1 inhibitor [2] and by the expression of a cysteine protease inhibitor (AtCYS1) [25]. Although studies report caspase activity in plants [26–28], and transgenic plants with an overexpression of a caspase inhibitor (protein p35 and Op-IAP) show the inhibition of HR [29, 30], thus far no analogue of animal caspases has been found in plants. It is rather metacaspases, that is, functional homologues of mammalian capases, that are present in plant tissue [31]. Belenghi *et al.* [32] showed that *Arabidopsis thaliana* metacaspase 9 (AtMC9) may be kept inactive through *S*-nitrosylation of a critical cysteine residue of the AtMC9. In turn, the mature form of this executor of cell death is insensitive to *S*-nitrosylation. It is possible that NO, through *S*-nitrosylation, could help maintain plant proteases such as metacaspases in their inactive state by preventing autocleavage and cognate activation [33].

It has been attempted to explain the interplay between ROS and NO during PCD establishment and regulation with the use of different models [34, 35]. It may be assumed that between free radicals, generated upon pathogen challenge, there are complex relations, dependent on quantitative ratios and physiological conditions of the plant, which determine their effectiveness in the stimulation of PCD and defense reactions, for example, via intracellular signals [5, 34, 36]. According to Delledonne *et al.* [34], only a tightly balanced production of NO and ROS (mainly H_2O_2) triggers the mechanism of cell death. In soybean and tobacco cell suspensions, a simultaneous increase in NO and H_2O_2 activated cell death, whereas an independent increase in only one of the above-mentioned factors induced cell death only slightly [1, 37]. Moreover, cytological observations showed that either the administration of NO donors or a change in the H_2O_2 level had no effect on elicitation of HR in oat cells infected with *P. coronata* f. sp. *avenae*, although both molecules were required for the onset of death in neighboring cells [4]. The mechanism by which NO and H_2O_2 interact in killing cells is still largely unknown. From the chemical point of view, NO and H_2O_2 may form singlet oxygen or hydroxyl radicals and through these molecules can cause cell death [38]. In addition, both NO and H_2O_2 at higher concentrations may provoke lipid peroxidation and DNA damage [39]. *In planta* NO and H_2O_2-mediated cell death appears to be triggered by the stimulation of active processes of death with the modulation of the caspase-like signaling cascade and release of cytochrome *c* from mitochondria [40]. Experiments using pharmacological and genetic approaches provide additional evidence for a partnership between NO and H_2O_2 during PCD induction [41]. In this case, catalase-deficient (CAT1AS) tobacco plants, treated with exogenous NO and exposed to a moderate high light, exhibited strong potentiation of cell death compared to wild-type plants. Moreover, studies conducted on a transgenic line of *A. thaliana* with an overexpression of the thylacoidal form of APX showed reduced symptoms of cell death when infiltrated with the NO-donor SNP [42]. In contrast, *Arabidopsis* tAPX antisense plants showed enhanced symptoms of NO-induced cell death, which confirms that both H_2O_2 and

NO participate in triggering cell death by oxidative stress [43]. In cells subjected to HR, the regulation of a functional H_2O_2 concentration seems to occur by a reversible inhibition of both APX and CAT [44, 45]. A decrease in APX activity, with simultaneous generation of H_2O_2 and NO during the induction of PCD, was observed by de Pinto *et al.* [37]. In non-*R*-gene resistance of pelargonium leaves, transient suppression of APX and CAT via NO correlated with enhanced H_2O_2 accumulation was observed in both resistant and susceptible genotypes [15]. A decrease in activity in the case of both CAT and APX may be explained by NO binding to the heme centers of these enzymes [2]. In pharmacological approaches using NO donors, the duration of this inhibition is combined with the time of decomposition of the particular donor compound [46].

According to Zaninotto *et al.* [38], relative rates of ROS production are pivotal in modulating NO reactivity. If the NO/$O_2^{\bullet-}$ balance is in favor of $O_2^{\bullet-}$, NO is scavenged before it can react with H_2O_2. In contrast, if the NO/$O_2^{\bullet-}$ balance is in favor of NO, $O_2^{\bullet-}$ is inactivated before it is dismutated to H_2O_2. Both options lead to the formation of the peroxynitrite anion ($ONOO^-$), whereas the cooperation of NO and H_2O_2 leads to plant cell death [34].

Recent findings by Zhao *et al.* [35] showed that both NO and $O_2^{\bullet-}$ were required for elicitor-induced cell death in *Cupressus lusitanica* cell culture, whereas H_2O_2 did not significantly influence cell death. NO and H_2O_2 enhanced each other's production, but NO production negatively affected $O_2^{\bullet-}$ accumulation. A direct interaction between NO and $O_2^{\bullet-}$ may be implicated in the production of a potent oxidant, peroxynitrite, which might mediate elicitor-induced cell death. In animal systems, peroxynitrite is a highly toxic molecule to cells, mediating apoptosis, capable of translocation at short distances, and reaching subcellular compartments and neighboring cells [47]. For plants, $ONOO^-$ seems to be relatively nontoxic [34]. However, it was found that in tobacco leaves, $ONOO^-$ induces PR-1 accumulation [20] and protein nitration modulating the cell redox state [34]. According to Almillo and Garcia-Olmedo [48], direct application of $ONOO^-$ to plants induced cell death, which was not observed when urea (an $ONOO^-$ scavenger) was added. Furthermore, using the capacity of urea to trap $ONOO^-$, it was shown that although peroxynitrite was responsible for death of most *Arabidopsis* cells in response to avirulent *P. syringae*, scavenging of this anion did not lead to an effective defense against avirulent bacteria [48].

NO, after being transformed into a peroxynitrite ion, may cooperate in killing microorganisms [20, 49]; however, it has not been clarified thus far whether NO and its derivatives are directly toxic to pathogens in plants [50]. It was demonstrated *in vitro* that growth of virulent and avirulent strains of *Pseudomonas* was inhibited by both NO and the system generating peroxynitrite (SNP + hypoxanthine/xanthine oxidase) [51, 52]. Romero-Puertas *et al.* [49] suggested that $ONOO^-$ may be continuously formed in healthy cells, so plants may develop detoxification mechanisms. In animal cells, ascorbates may play a significant role in inactivating $ONOO^-$ [53]. Taking into consideration the fact that ascorbic acid (AsA) is a quantitatively dominant antioxidant in plant cells [54], it is possible that AsA participates in $ONOO^-$ decomposition also in plant cells.

The presence of NO seems essential in cells undergoing PCD; however, there is evidence that no accumulation of NO was required before cell collapse [3]. According to these authors, in *Arabidopsis* leaves infected with *P. syringae* pv. tomato, NO generation did not occur sufficiently early for NO to be a signaling component controlling HR triggering, which suggests the role of an intercellular signal in cell-to-cell spread of HR. In turn, Mur *et al.* [10], using the laser photoacoustic technique for NO measurement, detected NO within 40 min of the challenge with *P. syringae* pv. *phaseolicola*, some 5 h before the initiation of visible tissue collapse.

Summing up, the cooperation of NO and ROS in triggering and execution of cell death is very complex and still enigmatic. In addition to the effect of NO and ROS on HR induction, these highly reactive molecules also participate in other plant defense responses, for example, in rebuilding and lignification of the plant cell wall, limiting pathogen spread over the host tissue.

10.4 Cross Talk of NO with Salicylic Acid, Jasmonic Acid, and Ethylene

Plants are capable of differentially activating distinct defense pathways, depending on the type of invader feed preference [55]. Plant signaling molecules such as salicylic acid (SA), jasmonic acid (JA), and ethylene play an important role in the intercellular network, and cross talk between them is thought to provide a significant regulatory potential for activating multiple resistance responses in varying combinations. Moreover, it may help the plant to prioritize the stimulation of a particular defense mechanism over another [56]. In some cases, it has been shown that SA is a potent inhibitor of JA and ethylene-dependent defense responses [57]. There is also evidence for the inhibition of salicylate action by JA [58]. In nature, however, plants often deal with simultaneous or sequential invasions by multiple aggressors, which can modify the primary induced defense response of the host plant [59]. Thus, a question arises as to how NO cooperates with SA, JA, and ethylene in fine-tuning multifarious defense responses. Evidence obtained so far demonstrates the involvement of and interaction between NO, JA, SA, and ethylene during cell response in a highly complex network. Synergistic and antagonistic actions between these signals have been observed, depending on the type of biotic stress, the strength of its effect, and the specificity of the plant–pathogen system.

It is well documented that NO acts through the SA signaling pathway [60]. Nitric oxide treatment of tobacco leaves induced a significant increase in endogenous SA, which was required for the expression of defense genes such as *PR-1* and *PAL*. Subsequently, using transgenic NahG tobacco, it was shown that SA was required for the NO-mediated induction of SIPK and concluded that SIPK may function downstream of SA in the NO signaling cascade for defense responses [61]. On the other hand, NO functionally requires SA, which is probably engaged in the potentiation of NO effects in eliciting innate resistance [62] and in the establishment of systemic acquired resistance [20]. Although SA is a key defense molecule in both local and systemic resistance, it is not equally effective against all pathogenic microorganisms of a plant.

Jasmonic acid is another plant signaling molecule involved in defense responses to necrotrophic pathogens and insects. NO may negatively modulate wound response in tomato [63], which could be linked to the JA-inducible expression of arginase, which in turn antagonizes NO production [64]. However, Huang *et al.* [65] reported that in *Arabidopsis*, local injury leads to prompt NO accumulation, which induces key enzymes of JA biosynthesis. Also, the results reported by Xu *et al.* [66] indicated an interdependence between NO generation induced by a fungal elicitor from *Aspergillum niger*, JA biosynthesis, and the production of hypericin in *Hypericium perforatum* cells. Treating *H. perforatum* cells with an NO scavenger, that is, cPTIO, and JA biosynthesis inhibitors inhibits not only the NO production and JA accumulation induced by the elicitor but also the production of hypericin. Moreover, the inhibitor of NOS and cPTIO stopped elicitor-induced NO and JA biosynthesis, while JA synthesis inhibitors did not reverse NO generation. These events suggest that JA functions as a second messenger of NO and its production is regulated by NO [66].

NO may also influence ethylene biosynthesis since the application of exogenous NO to plants modulates the generation of ethylene [67–70]. A study by Lindermayr *et al.* [69] showed that NO directly acts by downregulating ethylene synthesis through *S*-nitrosylation of methionine adenosyltransferase (MAT1) in *Arabidopsis* plants. The attachment of NO leads to the inhibition of MAT1 activity and results in depletion of the pool of an ethylene precursor *S*-adenosylmethionine (SAM). A decrease in ethylene production synchronized with a reduction of disease symptoms via NO showed the opposite role of both signals in pelargonium resistance to *B. cinerea* [15]. Evidence of the interplay between NO and ethylene in the senescence of plant tissues suggests an antagonistic effect of both molecules, found previously by Leshem *et al.* [71]. The possibility that cross talk between NO and ethylene also plays a role in disease suppression is intriguing because ethylene is required for the timing of the onset of senescence and therefore may be involved in the link between pathogenesis and senescence-dependent PCD. Although this idea seems generally to be true, the actual situation is rather more complicated. Recently, Mur *et al.* [72] found that NO and SA were required to generate the biphasic pattern of ethylene production in tobacco, following inoculation with the HR-eliciting avirulent strain of *P. syringae*.

It must be stressed that NO and the signals discussed seem to either synergize or antagonize in their signaling functions at different concentrations and under changing physiological conditions. Collectively, these data suggest that the interplay between these signals mediates a variety of defense responses. Results presented by Zago *et al.* [41] indicated that NO may act either individually or in partnership with H_2O_2 in a complicated cross talk with all hormones and other signals in regulating defense gene expression and in the execution of HR.

10.5 The Role of NO in the Micro- and Macroscale of Plant Communication

In contrast to other abiotic stressors, only the attack of pathogenic microorganisms leads to an increased NO synthesis limited to the infection site. Point necroses, frequently found on plants, are signs of death of single host cells as a consequence of infestation by viruses or bacteria.

10.5.1
NO Cell Signaling Domain

If the plant recognizes an invader, a rapid (i.e., within the first few minutes) and transient induction of NO synthesis occurs in different compartments of the challenged cell. Similar sites of NO synthesis and accumulation were described by Neill *et al.* [73] as "NO hot-spots" or NO signaling microdomains. NO in each cell may be generated both from nitrite by cytosolic and membrane-bound nitrate reductase activity and from the arginine-dependent pathway via nitric oxide synthase-like activity and other mechanisms of NO biosynthesis localized in the mitochondria, peroxisomes, plastids, and apoplast spaces. However, overproduction of NO may provoke disrepair effects in challenged cells (e.g., induction of HR) and, after all, is beneficial from the viewpoint of the fitness of the whole plant. Biological responses via NO depend on the gradient of local and distal NO concentrations as an effect of the rate of synthesis, translocation, and removal of this reactive nitrogen species. NO has also the ability to direct interaction with target molecules and other signals traveling and amplifying its effects. A rapid induction of defense responses by the resistant plant through NO at the site of invader recognition might rely on controlling free Ca^{2+} mobilization from intracellular stores, via activation of Ca^{2+}-permeable channels related to RYRs [11], promoting *S*-nitrosylation of redox-related and metabolic proteins and an uncontrolled intensification of protein tyrosine nitration [74]. NO, after being transformed into a peroxynitrite ion, may cooperate in killing microorganisms [20, 49] and mediate in orchestration with H_2O_2 in the elicitation of cell death [41].

It must be remembered, however, that induction of HR is usually associated with triggering and propagation of SOS signals through a complex network and addressed to noninfected cells.

10.5.2
NO in Short-Distance Communication

From the microdomain of enhanced NO synthesis restricted to the target cell attacked by a pathogen, NO may penetrate surrounding cells and trigger other signals of cell-to-cell communication, for example, H_2O_2, SA, or ethylene, and finally induce local defense responses [75]. Nitric oxide is a gaseous free radical with a relatively long (compared to other free radicals) half-life, that is, 3–5 s in biological systems [76, 77]. It is one of the smallest diatomic molecules with a high diffusivity ($4.8 \times 10^{-5}\, cm^2\, s^{-1}$ in H_2O), exhibiting hydrophobic properties. Thus, NO may not only easily migrate within the hydrophilic regions of the cell such as the cytoplasm but also freely diffuse through the lipid phase of membranes at a diffusion distance of probably no more than 500 μm [78] with a speed of migration of about $50\, \mu m\, s^{-1}$.

Using two low-invasive techniques of NO *in situ* measurement, an electrochemical and a cytochemical method with fluorochrome, it has been found in our previous study [15] that NO burst, although genetically conditioned, was not restricted only to the *avr–R* interaction. The attack of the necrotrophic pathogen, *B. cinerea*, on leaves of a resistant pelargonium genotype generated a sufficiently strong NO burst to induce periodical changes in the signaling network, leading to H_2O_2 upregulation, ethylene downregulation, and triggered HR of challenged cells, revealed as TUNEL-positive cell death. The bioimaging with DAF-2DA showed that NO synthesis expanded from the epicenter to neighboring cells and then gradually disappeared during successive hours. The following wave of secondary NO generation was able to fine-tune resistance responses against the necrotrophic fungal pathogen by provoking noncell death-associated resistance (TUNEL-negative) with an enhanced pool of antioxidants, which finally favored the maintenance of homeostasis of surrounding cells.

Very rapidly or slowly, but adequate for invader recognition, the host plant is able to create physical barriers, synthesize PR proteins and other antimicrobial compounds. These defense responses are usually triggered in the zone of surrounding cells, neighboring those target cells challenged by the pathogen.

Phytoalexin accumulation is one of the events associated with short-distance transport of defense signals, in which NO seems to be engaged [79]. Treating potato tubers with exogenous NO stimulated the accumulation of rishitin, synthesized after *P. infestans* infection. In addition, the effect of inhibited production of this compound was observed after application of an NO scavenger [80]. Independently, NO induced the biosynthesis of specific phytoalexins such as isoflavonoids and pterocarpons in soybean cotyledons in response to the elicitor of *Diaporthe phaseolurum* f. sp. *meridionalis* [81], and catharantine production in *Catharantus roseus* cell cultures [82, 83]. Furthermore, β-thujaplicin production associated with NO synthesis and HR induction was recently found by Zhao *et al.* [35] in *C. lusitanica* cells treated with yeast elicitor culture.

10.5.3
NO from Cross- to Long-Distance Communication

NO may also participate in systemic defense signaling, leading to systemic acquired resistance. In tobacco, exogenous NO induces SA accumulation, playing a fundamental role in SAR [60]. The application of inhibitors specific for animal NOS or NO scavengers reduced SAR [62]. These results suggest an important role of NO in the induction of a distal signaling network concomitant with SAR establishment in tobacco. In plants, similar to the cardiovascular system of mammals, NO may be transported in the form of nitrosoglutathione (GSNO) [20]. It is assumed that GSNO may function both as an intracellular and as an intercellular NO carrier and is transported throughout the plant via phloem over long distances. Furthermore, it was shown that GSNO induces systemic resistance against TMV in tobacco [62] and is an effective inducer of PAL as well [60]. Recently, much attention has been focused on nitrosoglutathione reductase (GSNOR) that metabolizes and regulates *S*-nitrosothiol levels [84]. Mutation within *Arabidopsis* GSNOR elevated a cellular *S*-nitrosothiol pool; however, this increased disease susceptibility was associated with a reduction of SA accumulation and suppression of SA-dependent genes, for example, PR-1 [85]. In contrast, a positive correlation between *S*-nitrosothiol levels and plant defense against pathogens was recently documented by Rustérucci *et al.* [86].

Another candidate that could serve as a mobile signal responsible for NO involves conjugates with polyamines (PAs). Such a direct chemical link between NO and PAs forming a NONOates bond may move through the phloem and serve the role of a long-distance signal; however, this hypothesis calls for experimental verification [87].

Recently, it was found by Gaupels *et al.* [88] that NO was induced in the phloem system of *Vicia faba* by H_2O_2 and SA local application, while phloem exudates arising from H_2O_2-treated *Cucurbita maxima* contained an increased level of nitrated proteins. NO, H_2O_2, and SA-mediated defense signaling partially overlaps, since the SA positive feedback loop is essential for amplifying the distal signal in the upper zone of the plant [89, 90].

Figure 10.2 summarizes nitric oxide signaling in a model encompassing three NO domains on the microscale (cell, tissue) and macroscale (organ) of the plant, which together may be an integral part of the signaling network leading to SAR establishment.

10.6 Does NO Participate in Stressful Memory of the Plant?

As mentioned above, plants, following a primary stress, occasionally acquire resistance to a much stronger secondary stress. So far, much experimental evidence has been collected indicating the functioning in a plant, exposed to adverse environmental factors, of a certain type of "memory" or molecular imprint of an experienced primary stress, which facilitates its rapid and effective response when a secondary stress occurs [91].

This phenomenon, in relation to abiotic stresses, is termed acclimation or hardening, whereas in relation to biotic stresses it is usually described as acquired SAR-type or ISR-type (induced systemic resistance) mechanisms. Enhanced stress responses of plants are generally the result of the so-called priming with the same stressor (different pathogenic microorganisms) or a chemical inducer.

In spite of intensive studies conducted for many years, no fully satisfactory results have been reported that would make it possible to clarify the nature of this resistance and the manner in which it is realized. In this respect, it seems crucial to explain the mechanism of imprinting and duration of the readiness state, that is, changes in defense predisposition of the plant. Among endogenous signals, which potentially modulate metabolic responses in the course of resistance acquisition, a special role is played by nitric oxide. This important molecule, the synthesis of which increases rapidly during stress, may both cooperate with other signals and intracellular hormones and participate in the transduction of stress information, for example, via *S*-nitrosylation/denitrosylation of metabolic, structural, and signaling proteins and protein transcription factors.

A reversible process of protein nitrosylation through NO may serve similar functions in the signal transduction pathway as phosphorylation/dephosphorylation reactions. Lindemayr *et al.* [92] identified 52 *S*-nitrosylated proteins, including stress-related, redox-related, signaling or regulatory, cytoskeletal, and metabolic polypeptides, from *A. thaliana* leaves after NO-donor application. Little is known about NO influence on the transcriptional factors in the plant. Serpa *et al.* [93] first

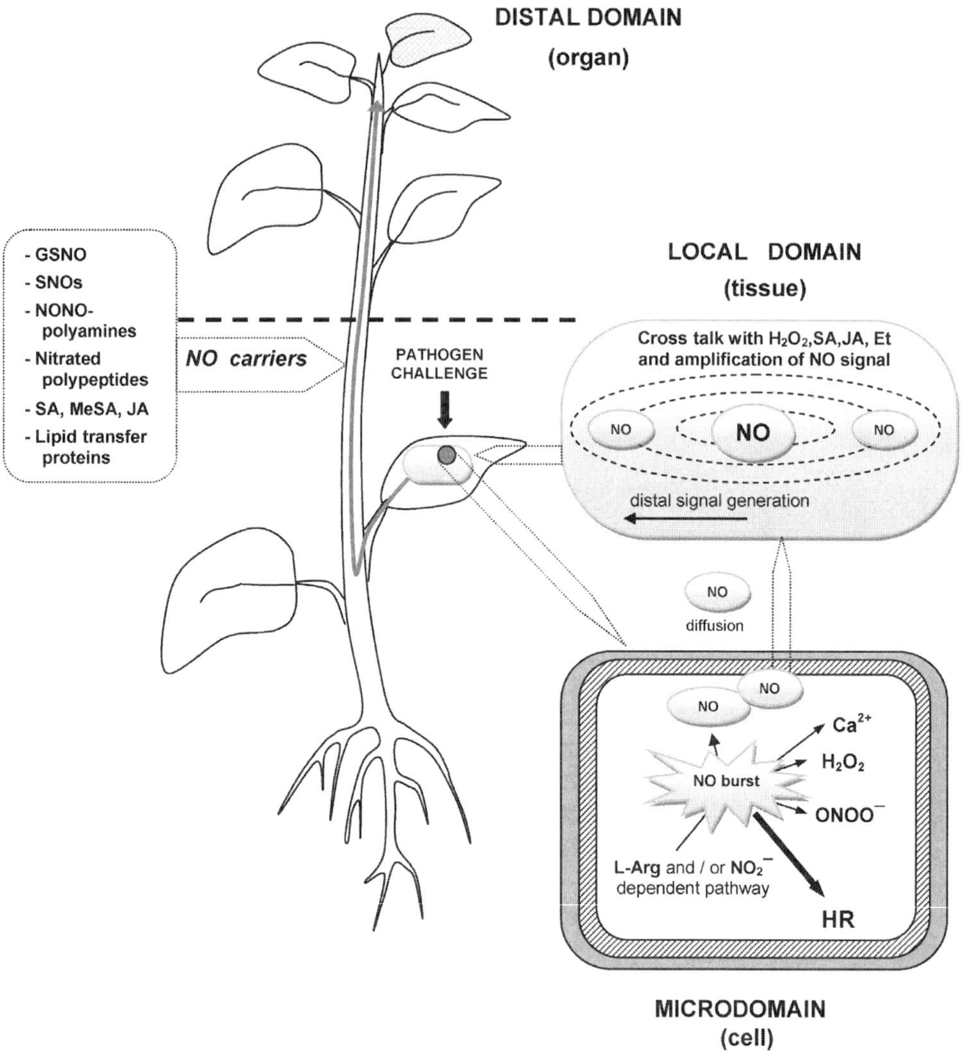

Figure 10.2 Nitric oxide signaling in a model encompassing three NO domains on the microscale (cell, tissue) and macroscale (organ) of the plant that together may be an integral part of signaling network leading to SAR establishment. *NO microdomain* – attacked by the invader, target cell provokes an immediate NO burst. NO governs the control of free Ca^{2+} mobilization from intracellular stores and with H_2O_2 and/or peroxynitrite ion induces host cell death (HR) or kills the pathogen. *NO local domain* – NO may diffuse to noninfected cells and together with the local signaling network of cell-to-cell communication, for example, H_2O_2, SA, or ethylene, may amplify signal, induce local defense responses, and generate mobile resistance signals. *NO distal domain* – NO may also participate in systemic defense signaling, leading to SAR. Candidates that could serve as mobile signals responsible for NO may be GSNO (nitrosoglutathion), SNOs (nitrosothiols), NONO-polyamines (NO-PA conjugates), nitrated proteins, hormones (JA, MeSA), and lipid transfer proteins.

established that NO may inhibit DNA-binding activity of AtMYB2 from *A. thaliana* by posttranscriptional *S*-nitrosylation of its conserved Cys53 residue. More than 125 members of the R2R3-MYB family of plant transcription factors, being potential targets for NO covalent modification, regulate different metabolic mechanisms, including responses to environmental factors and hormones [94].

The capacity to obtain and maintain an efficient pattern of gene expression seems key to every organism exposed to environmental stresses. The present state of knowledge is insufficient to fully answer the question concerning the way in which information on experienced primary stress is encoded. According to the concept proposed by Conrath *et al.* [95], in the state of enhanced defense readiness, induced accumulation of key transcription factors takes place, which might facilitate the initiation of rapid expression of defense genes following a secondary infection. However, it is still not clear whether there is any (and if yes, *what*) role, which NO may have in stress imprint formation, potentially consisting of both epigenetic changes and accumulation of modified proteins and transcription factors.

10.7 NO and Plant Recovery from Stress

Among various environmental stresses, wounding seems to be the most crucial, being a potential entry point for invading pathogens and simultaneously inhibiting normal growth and plant reproduction [96]. The capability of plants to recover after stress depends on coordinated responses, which are induced or accelerated between a few minutes to several hours after tissue damage [97]. The response involves the generation and transduction of multiple signals, subsequently leading to the expression of wound-responsive genes. Many structurally different molecules are associated with the wound-signaling network including jasmonates, salicylic acid, systemin, oligosaccharides, abscisic acid, and ethylene [97]. Recent studies have revealed that NO might also be involved in tissue injury and promote wound-healing responses [23, 65, 98].

In epidermal cells of *Arabidopsis*, a wound-induced NO burst was triggered within several minutes, which was confirmed by the use of a fluorochrome, DAF-2DA [65]. Similarly, as early as 5 min after cutting at the site of the injury, a 2-h NO generation was recorded in epidermal cells of pelargonium leaves using an NO-selective microelectrode and bioimaging with DAF-2DA [99].

When a plant is slightly wounded, it responds quickly by restoring destroyed tissues and protecting them from the subsequent influence of environmental insults, including pathogen attack. In animal systems, NO is synthesized during tissue wound healing, and inhibition of its production impairs the recovery phase [100]; however, a correlation between NO and the outcome of healing is still unclear. It is important to note that L-arginine – the substrate for NO synthesis – may also be metabolized in wounds via arginase activity to ornithine, being a precursor of proline and polyamine synthesis [101]. In animals, this may be a significant element in the wound-healing pathway, since proline serves as a substrate for collagen synthesis, whereas polyamines are involved in cell proliferation [102, 103]. In plants, little is known whether NO can directly affect wound healing. However, the presence of hydroxyproline components in the wound signal – systemin [104] and stress-dependent oxidative cross-linking of cell wall structural proline-rich (GRP) and hydroxyproline-rich proteins (HRGP) – plays an important role in cell protection [105]. Recent findings confirm that NO probably promotes the wound-healing response of plant tissue, leading to a reinforcement of the cell wall [99, 106].

Using SNP as an NO donor, Paris *et al.* [106] showed induction of callose deposition and an increase of transcript levels of two wounding markers, extensin and PAL, in the damaged area of potato leaflets. In addition, wounded NO-donor-treated leaves showed better tissue integrity and sustained higher chlorophyll levels. In nicked pelargonium leaves, an early NO generation, accompanied by a transient H_2O_2 accumulation, was involved in callose deposition as well as in lignin formation. It is proposed that NO plays a buffering role in the maintenance of cell homeostasis in leaves subjected to injury through, for example, a restriction of the depletion of the low molecular weight antioxidant pool (i.e., ascorbic acid and thiols). A transient increase in the level of endogenous H_2O_2, observed in pelargonium, was linked to a temporary inhibition of CAT and APX activity via NO. Hydrogen peroxide produced after injury is thought to diffuse into disrupted cells and activate many of the plant responses postwounding, such as sealing and reconstruction changes of damaged tissue through lignin formation and callose deposition [99].

It is possible that NO functions as a modulator of plant-wounding responses through an amplification of other regulatory mechanisms. Its cross talk with JA and SA signaling pathways particularly seems to be of interest. It was found in wild-type *Arabidopsis* plants that exogenous NO strongly induces SA accumulation and key enzymes of jasmonic acid biosynthesis such as allene oxide synthase (AOS) and lipoxygenase (LOX2); however, it did not induce JA and accumulation of typical JA-inducible genes (PDF1.2, JIP). Moreover, wound-induced AOS gene expression was NO-independent [65]. On the other hand, NO treatment of SA-deficient NahG plants resulted in the activation of JA-responsive genes as well as in an increase in JA level, suggesting a role for SA in the suppression of JA- or NO-responsive genes [65]. In tomato, NO inhibited wound-inducible H_2O_2 generation and proteinase inhibitor gene expression downstream of jasmonic acid synthesis, and it has been shown that NO has a role in downregulating the expression of wound-inducible defense genes during pathogenesis [63].

Pathogen infection or insect infestation is usually accompanied by tissue damage, although the amount of NO generated in response to a pathogenic factor seems to considerably exceed the level recorded during leaf injury. For example, a fungal elicitor, cryptogein, caused more than a fourfold increase in NO emission in epidermal cells of tobacco leaves [5]. According to Huang *et al.* [65], the kinetics of NO synthesis in wounded *Arabidopsis* leaf is similar to an elicitor-induced NO generation in tobacco. In turn, in pelargonium leaf postwounding NO generation, measured by an electrochemical technique *in situ*, was almost 10-fold weaker compared to that in the resistant pelargonium challenged with *B. cinerea* [15]. It might justify an opinion that the intensity, duration, and kinetics of NO generation include information about the type of stressor being recognized by the plant.

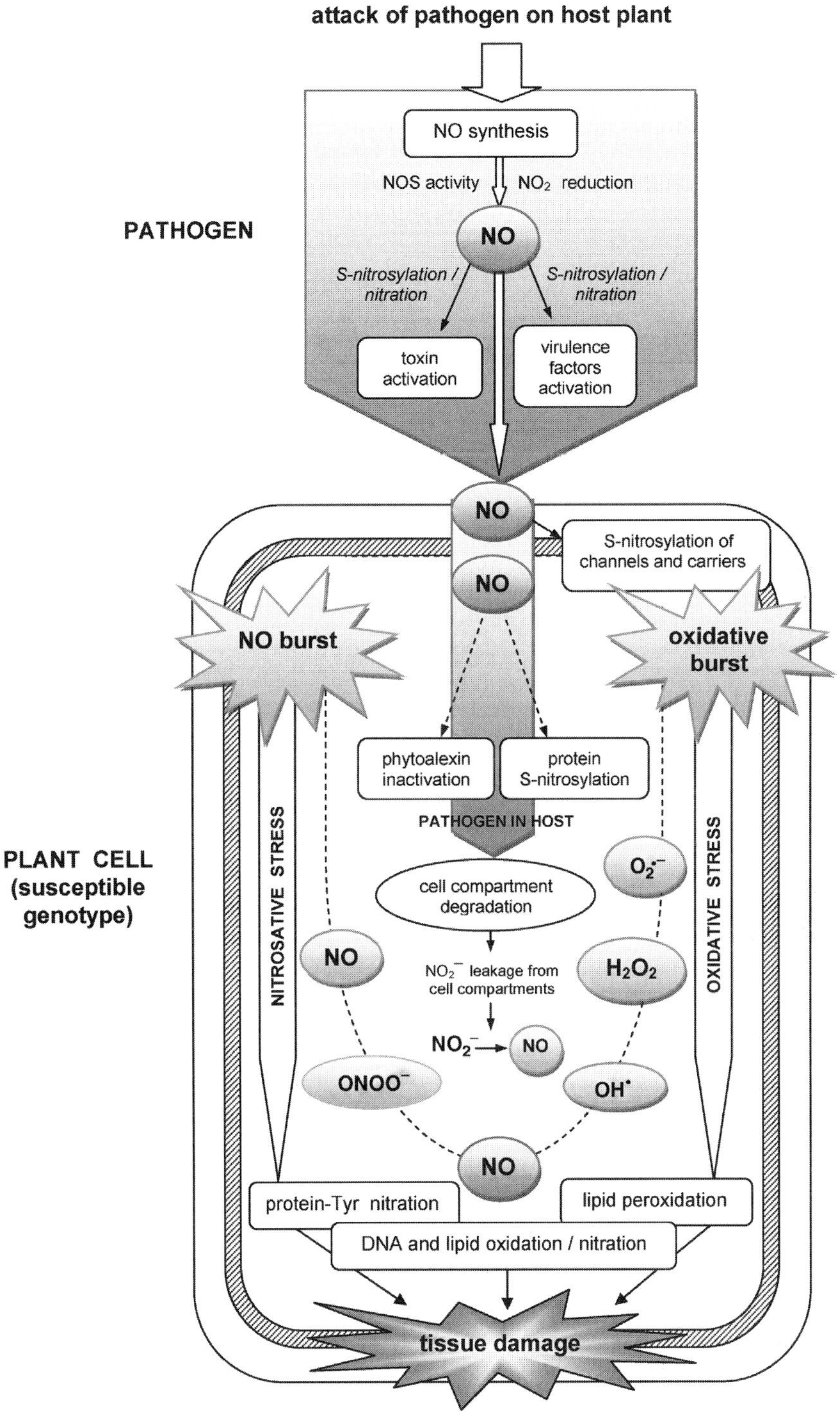
attack of pathogen on host plant
PATHOGEN
NO synthesis
NOS activity
NO2 reduction
NO
S-nitrosylation / nitration
S-nitrosylation / nitration
toxin activation
virulence factors activation
NO
S-nitrosylation of channels and carriers
NO
NO burst
oxidative burst
phytoalexin inactivation
protein S-nitrosylation
PATHOGEN IN HOST
PLANT CELL (susceptible genotype)
NITROSATIVE STRESS
OXIDATIVE STRESS
cell compartment degradation
O2·−
NO
NO2− leakage from cell compartments
H2O2
NO2− → NO
ONOO−
OH·
NO
protein-Tyr nitration
lipid peroxidation
DNA and lipid oxidation / nitration
tissue damage

10.8 NO in the Offensive Strategy of the Pathogen

A conflict of interest has been observed between pathogens causing diseases and plants as potential objects of their attack. In the course of evolutionary change, microorganisms parasitizing plants have developed highly effective offensive strategies. So far, the role of NO in disease resistance, as well as in human, animal, and plant tissues, was mainly analyzed from the viewpoint of the host organism. Little is known about NO generation by the pathogenic microorganism. However, it is suspected that NO may be an effective weapon used by pathogens. In endothelial cells, pathogenic microbes and viruses may induce *S*-nitrosylation of membrane receptors to facilitate cellular entry [107]. There is evidence that NO generated by microorganisms may be responsible for the activation/deactivation of phytotoxins. In *Streptomyces* spp. nitric oxide, synthesized via NOS, nitrates taxtomine A, which determines the phytotoxicity of this compound [108, 109]. Moreover, nitration of lipopeptide arylomycin generated by *Streptomyces* sp. Tü 6075 enhances its antibacterial activity [110], which may be significant in the infestation of new ecological niches by this bacteria. NO has been shown to activate various antioxidant genes in *Escherichia coli* and *Bacillus subtilis* to protect microbial cells against oxidative and nitrosative stress [111].

So far, there are few studies on the mechanism and site of NO synthesis in pathogenic microorganisms challenging the plant. Wang and Higgins [112] recently reported on the accumulation of NO in conidia and germinating hyphae of *Colletotrichum coccodes*, in which intensity and location changed, depending on the developmental stage of the fungus. Also, Conrath *et al.* [8], using mass spectrometry (MIMS/RIMS), demonstrated the ability of pathogenic fungi, that is, *Phytium*, *Botrytis*, and *Fusarium* spp., to produce nitrite-induced NO.

When a necrotizing pathogen colonizes plant tissues, it destroys intracellular compartments of the host that contributes to uncontrolled NO overproduction from nitrite diffused from the chloroplasts. In our experiments, the necrotrophic invader *B. cinerea*, spreading over leaf tissue of the susceptible genotype, also produced huge quantities of NO and enhanced nitrosative and oxidative stress in the host plant [15].

Figure 10.3 presents the main effects of NO production derived from the invading pathogen on a susceptible genotype of the plant host during disease development.

Figure 10.3 Effects of NO production derived from invading pathogen on susceptible genotype of plant host during disease development. NO generated via NOS and/or nitrite pathway by microorganisms may be responsible for the activation of their phytotoxins and virulence factors. The putative pathogen is probably able to induce *S*-nitrosylation of host membrane receptors and/or channels to facilitate cellular entry. When the invader colonized plant tissues, it destroyed intracellular compartments of the host, which contributed to uncontrolled NO overproduction from nitrite diffused from the chloroplasts. Such huge quantities of NO in the susceptible genotype enhanced nitrosative and oxidative stress, leading to uncontrolled tyrosine nitration, lipid peroxidation, and tissue damage.

Little information exists regarding nitrosative stress during pathogenesis in plants. However, overproduction of reactive nitrogen species and protein tyrosine nitration have been found to occur in the olive leaves during salinity [113]. In animal cells, an increase in tyrosine nitration is rather the marker of stress severity, although other assignments of the 3-nitrotyrosine compounds should not be discarded [114].

10.9 Concluding Remarks

In recent years, significant progress has been made in elucidating the molecular mechanism underlying the interplay between NO and other hormone-regulated defense signaling pathways. However, it is well known that NO in orchestration with other players of the signaling network plays a key role during plant–pathogen interaction, by triggering resistance-associated cell death and inducing local and distal defense-related responses. Obviously, this cross talk between NO and plant hormones is sometimes puzzling and requires additional research. Moreover, it may be anticipated that in-depth information on the sources of NO synthesis in the pathogen and advances in the recognition of the role of NO in invader metabolism may provide new tools for a rational crop control to improve plant disease resistance.

References

1 Delledonne, M., Xia, Y., Dixon, R.A., and Lamb, C. (1998) *Nature*, **394**, 585–588.

2 Clarke, A., Desikan, R., Hurst, R.D., Hancock, J.T., and Neill, S.T. (2000) *Plant J.*, **4**, 667–677.

3 Zhang, C., Czymmek, K.J., and Shapiro, A.D. (2003) *Mol. Plant–Microbe Interact.*, **16**, 962–972.

4 Tada, Y., Mori, T., Shinogi, T., Yao, N., Takahashi, S., Betsuyaku, S. *et al.* (2004) *Mol. Plant–Microbe Interact.*, **17**, 245–253.

5 Foissner, I.D., Wendehenne, D., Langebartels, C., and Durner, J. (2000) *Plant J.*, **23**, 817–824.

6 Huang, J., Sommer, E.M., Kim-Shapiro, D.B., and King, S.B. (2002) *J. Am. Chem. Soc.*, **124**, 3473–3480.

7 Yamamoto, A., Katou, S., Yoshioka, H., Doke, N., and Kawakita, K. (2003) *J. Gen. Plant Pathol.*, **69**, 218–229.

8 Conrath, U., Amoroso, G., Kohle, H., and Sultemeyer, D.F. (2004) *Plant J.*, **38**, 1015–1022.

9 Guo, P., Cao, Y., Li, Z., and Zhao, B. (2004) *Plant Cell Environ.*, **27**, 473–477.

10 Mur, L.A.J., Santosa, I.E., Laarhoven, L.J.J., Holton, N.J., Harren, F.J., and Smith, A.R. (2005) *Plant Physiol.*, **138**, 1247–1258.

11 Lamotte, O., Gould, K., Lecourieux, D., Sequeira-Legrand, A., Lebrun-Garcia, A., Durner, J., Pugin, A., and Wendehenne, D. (2004) *Plant Physiol.*, **135**, 516–530.

12 Zeier, J., Delledonne, M., Mishina, T., Severi, E., Sonoda, M., and Lamb, C. (2004) *Plant Physiol.*, **136**, 2875–2886.

13 Delledonne, M. (2005) *Curr. Opin. Plant Biol.*, **8**, 390–396.

14 van Baarlen, P.V., Staats, M., and van Kan, J.A.L. (2004) *Mol. Plant Pathol.*, **6**, 559–574.

15 Floryszak-Wieczorek, J., Arasimowicz, M., Milczarek, G., Jeleń, H., and Jackowiak, H. (2007) *New Phytol.*, **175**, 718–730.

16 Levine, A., Tenhaken, R., Dixon, R., and Lamb, C. (1994) *Cell*, **79**, 583–593.

17 Govrin, E.M. and Levine, A. (2000) *Curr. Biol.*, **10**, 751–757.

18 Vleehouwers, V.G.A.A., van Dooijeweert, W., Govers, F., Kamoun, S., and Colon, L.T. (2000) *Planta*, **210**, 853–864.

19 Shetty, N.P., Lyngs Jörgensen, H.J., Due Jensen, J., Collinge, D.B., and Shekar Shetty, H. (2008) *Eur. J. Plant Pathol.*, **121**, 267–280.

20 Durner, J. and Klessig, D.F. (1999) *Curr. Opin. Plant Biol.*, **2**, 369–374.

21 Prats, E., Mur, L.A.J., Sanderson, R., and Carver, T.L.W. (2005) *Mol. Plant Pathol.*, **6**, 65–78.

22 Modolo, L.V., Augusto, O., Almeida, I.M.G., Pinto-Maglio, C.A.F., Oliveira, H.C., Seligman, K., and Salgado, I. (2006) *Plant Sci.*, **171**, 34–40.

23 Pedroso, M.C., Magalhaes, J.R., and Durzan, D. (2000) *J. Exp. Bot.*, **51**, 1027–1103.

24 Saviani, E.E., Orsi, C.H., Oliveira, J.F.P., Pinto-Maglio, C.A.F., and Salgado, I. (2002) *FEBS Lett.*, **510**, 136–140.

25 Belenghi, B., Acconcia, F., Trovato, M., Perazzolli, M., Bocedi, A., Polticelli, F., Ascenzi, P., and Delledonne, M. (2003) *Eur. J. Biochem.*, **270**, 2593–2604.

26 D'Silva, I., Poirier, G.G., and Heath, M.C. (1998) *Exp. Cell Res.*, **245**, 389–399.

27 Hatsugai, N., Kuroyanagi, M., Yamada, K., Meshi, T., Tsuda, S., Kondo, M., Nishimura, M., and Hara-Nishimura, I. (2004) *Science*, **305**, 855–858.

28 Rojo, E., Martin, R., Carter, C., Zouhar, J., Pan, S., Plotnikova, J., Jin, H., Paneque, M., Sanchez-Serrano, J.J., Baker, B., Ausubel, F.M., and Raikhel, N.V. (2004) *Curr. Biol.*, **14**, 1897–1906.

29 Dickmann, M.B., Park, Y.K., Oltersdorf, T., Li, W., Clemente, T., and French, R. (2001) *Proc. Natl. Acad. Sci. USA*, **98**, 6957–6962.

30 Del Pozo, O. and Lam, E. (2003) *Mol. Plant–Microbe Interact.*, **16**, 485–494.

31 Bozhkov, P.V., Suarez, M.F., Filonova, L.H., Daniel, G., Zamyatnin, A.A., Rodriguez-Nieto, S., Zhivotovsky, B., and Smertenko, A. (2005) *Proc. Natl. Acad. Sci. USA*, **102**, 14463–14468.

32 Belenghi, B., Romero-Puertas, M.C., Vercammen, D., Brackenier, A., Inzé, D., Delledonne, M., and Van Breusegem, F. (2007) *J. Biol. Chem.*, **282**, 1352–1358.

33 Hong, J.K., Yun, B.-W., Kang, J.-G., Raja, M.U., Kwon, E., Sorhagen, K., Chu, C., Wang, Y., and Loake, G.J. (2008) *J. Exp. Bot.*, **59**, 147–154.

34 Delledonne, M., Zeier, J., Marocco, A., and Lamb, C. (2001) *Proc. Natl. Acad. Sci. USA*, **98**, 13454–13459.

35 Zhao, J., Fujita, K., and Sakai, K. (2007) *New Phytol.*, **175**, 215–229.

36 Delledonne, M., Murgia, I., Ederle, D., Sbicego, P.F., Biondani, A., Polverari, A., and Lamb, C. (2002) *Plant Physiol. Biochem.*, **40**, 605–610.

37 de Pinto, M.C., Tomassi, F., and de Gara, L. (2002) *Plant Physiol.*, **130**, 689–708.

38 Zaninotto, F., la Camera, S., Polverari, A., and Delledonne, M. (2006) *Plant Physiol.*, **141**, 379–383.

39 Dubovskaya, L.V., Kolesneva, E.V., Knyazev, D.M., and Volotovskii, I.D. (2007) Protective role of nitric oxide during hydrogen peroxide-induced oxidative stress in tobacco plants. *Russ. J. Plant Physiol.*, **54**, 755–762.

40 Mur, L.A.J., Carver, T.L.W., and Prats, E. (2006) *J. Exp. Bot.*, **57**, 489–505.

41 Zago, E., Morsa, S., Dat, J.F., Alard, P., Ferrarini, A., Inze, D., Delledonne, M., and van Breusegem, F. (2006) *Plant Physiol.*, **141**, 404–411.

42 Murgia, I., de Pinto, M.C., Delledonne, M., Soave, C., and de Gara, L. (2004) *J. Plant Physiol.*, **161**, 777–783.

43 Tarantino, D., Vannini, C., Bracale, M., Campa, M., Soave, C., and Murgia, I. (2005) *Planta*, **221**, 757–765.

44 Mittler, R., Lam, E., Shulaev, V., and Cohen, M. (1999) *Plant Mol. Biol.*, **39**, 1025–1035.

45 Bestwick, C.S., Adam, A.L., Puri, N., and Mansfield, J.W. (2001) *Plant Sci.*, **161**, 497–506.

46 Floryszak-Wieczorek, J., Milczarek, G., Arasimowicz, M., and Ciszewski, A. (2006) *Planta*, **224**, 1363–1372.

47 Marla, S., Lee, J., and Groves, J.T. (1997) *Proc. Natl. Acad. Sci. USA*, **94**, 14243–14248.

48 Alamillo, J.M. and Garcia-Olmedo, F. (2001) *Plant J.*, **25**, 529–540.

49 Romero-Puertas, M.C., Perazzolli, M., Zago, E.D., and Delledonne, M. (2004) *Cell Microbiol.*, **6**, 795–803.

50 Garcia-Olmedo, F., Rodrigguez-Palenzulea, P., Molina, A., Alamillo, J.M., Lopez-Solanilla, E., Berrocal-Lobo, M., and Poza-Carrion, C. (2001) *FEBS Lett.*, **489**, 219–222.

51 Noronha-Dutra, A.A., Epperlein, M.M., and Woolf, N. (1993) *FEBS Lett.*, **321**, 59–62.

52 Garcia-Mata, C. and Lamattina, L. (2002) *Plant Physiol.*, **128**, 790–792.

53 Arteel, G.E., Briviba, K., and Sies, H. (1999) *FEBS Lett.*, **445**, 226–230.

54 Smirnoff, N. (2000) *Curr. Opin. Plant Biol.*, **3**, 229–235.

55 Thomma, B.P.H.J., Penninckx, I.A.M.A., Broekaert, W.F., and Cammue, B.P.A. (2001) *Curr. Opin. Immunol.*, **13**, 63–68.

56 Halim, V.A., Vess, A., Scheel, D., and Roshal, S. (2006) *Plant Biol.*, **8**, 307–313.

57 Glazebrook, J. (2005) *Annu. Rev. Phytopathol.*, **43**, 205–227.

58 Bostock, R.M. (2005) *Annu. Rev. Phytopathol.*, **43**, 545–580.

59 Van der Putten, W.H., Vet, L.E.M., Harvey, J.A., and Wackers, F.L. (2001) *Trends Ecol. Evol.*, **16**, 547–554.

60 Durner, J., Wendenhenne, D., and Klessig, D.F. (1998) *Proc. Natl. Acad. Sci. USA*, **95**, 10328–10333.

61 Kumar, D. and Klessig, D.F. (2000) *Mol. Plant–Microbe Interact.*, **13**, 347–351.

62 Song, F. and Goodman, R.M. (2001) *Mol. Plant–Microbe Interact.*, **12**, 1458–1462.

63 Orozco-Cardenas, M.L. and Ryan, C.A. (2002) *Plant Physiol.*, **130**, 487–493.

64 Chen, H., McCaig, B.C., Melotto, M., He, S.Y., and Howe, G.A. (2004) *J. Biol. Chem.*, **279**, 45998–46007.

65 Huang, X., Stettmaier, K., Michel, C., Hutzler, P., Mueller, M.J., and Durner, J. (2004) *Planta*, **218**, 938–946.

66 Xu, Y., Yuanlin, C., Tao, Y., and Zhao, B. (2005) *Meth. Enzymol.*, **396**, 84–92.

67 Leshem, Y.Y. and Haramaty, E. (1996) *J. Plant Physiol.*, **148**, 258–263.

68 Leshem, Y.Y. (2001) *Nitric Oxide in Plants: Occurence Function and Use*, Kluwer Academic Publishers, Dordrecht - Boston London.

69 Lindermayr, C., Saalbach, G., Bahnweg, G., and Durner, J. (2006) *J. Biol. Chem.*, **281**, 4285–4291.

70 Zhu, S.H. and Zhou, J. (2007) *Food Chem.*, **100**, 1517–1522.

71 Leshem, Y.Y., Wills, R.B.H., and Ku, V.V.V. (1998) *Plant Physiol. Biochem.*, **36**, 825–833.

72 Mur, L.A.J., Laarhoven, L.J.J., Harren, F.J.M., Hall, M.A., and Smith, A.R. (2008) *Plant Physiol.* doi: 10.1104/pp.108.124404

73 Neill, S., Bright, J., Desikan, R., Hancock, J., Harrison, J., and Wilson, I. (2008) *J. Exp. Bot.*, **59**, 25–35.

74 Stefano, M., Vandelle, E., Polverari, A., Ferrarini, A., and Delledonne, M. (2007) *Nitric Oxide in Plant Growth, Development and Stress Physiology* (eds L. Lamattina and J.C. Polacco), Springer, pp. 207–222.

75 Garcia-Brugger, A., Lamotte, O., Vandelle, C., Bourque, S., Lecourieux, D., Poinsot, B., Wendehenne, D., and Pugin, A. (2006) *Mol. Plant–Microbe Interact.*, **19**, 711–724.

76 Henry, Y.A., Ducastel, B., and Guissani, A. (1997) *Nitric Oxide Research from Chemistry to Biology* (eds Y.A. Henry, A. Guissani, and B. Ducastel), Landes Co. Biomed. Publ., Austin, TX, pp. 15–46.

77 Tuteja, N., Chandra, M., Tuteja, R., and Misra, M.K. (2004) *J. Biomed. Biotechnol.*, **4**, 227–237.

78 Yamasaki, H. (2005) *Plant Cell Environ.*, **28**, 78–84.

79 Able, A.J. (2003) *Protoplasma*, **221**, 137–143.

80 Noritake, T., Kawakita, K., and Doke, N. (1996) *Plant Cell Physiol.*, **37**, 113–116.

81 Modolo, L.V., Cunha, F.Q., Braga, M.R., and Salgado, I. (2002) *Plant Physiol.*, **130**, 1288–1297.

82 Zhao, J.A., Zhu, W.H., and Hu, Q. (2000) *Biotechnol. Lett.*, **22**, 509–514.

83 Zhao, J.A., Davis, L.C., and Verpoorte, R. (2005) *Biotechnol. Adv.*, **23**, 283–333.

84 Liu, L., Hausladen, A., Zeng, M., Que, L., Heltman, J., and Stamler, J.S. (2001) *Nature*, **410**, 490–494.

85 Feechan, A., Kwon, E., Yun, B.W., Pallas, J.A., and Loake, G.J. (2005) *Proc. Natl. Acad. Sci. USA*, **102**, 8054–8059.

86 Rustérucci, C., Espunya, M.C., Díaz, M., Chabannes, M., and Martínez, M.C. (2007) *Plant Physiol.*, **143**, 1282–1292.

87 Yamasaki, H. and Cohen, M.F. (2006) *Trends Plant Sci.*, **11**, 522–524.

88 Gaupels, F., Furch, A.C.U., Will, T., Mur, L.A.J., Kogel, K.-H., and van Bel, A.J.E. (2008) *New Phytol.*, **178**, 634–646.

89 Loake, G. and Grant, M. (2007) *Curr. Opin. Plant Biol.*, **10**, 466–472.

90 Vlot, A.C., Liu, P.P., Cameron, R.K., Park, S.W., Yang, Y., Kumar, D., Zhou, F., Padukkavidana, T., Gustafsson, C., Pichersky, E., and Klessig, D.F. (2008) *Plant J.*, **56**, 445–450.

91 Bruce, T.J.A., Matthes, M.C., Napier, J.A., and Pickett, J.A. (2007) *Plant Sci.*, **173**, 603–608.

92 Lindermayr, C., Saalbach, G., and Durner, J. (2005) *Plant Physiol.*, **137**, 921–930.

93 Serpa, V., Vernal, J., Lamattina, L., Grotewold, E., Cassia, R., and Terenzi, H. (2007) *Biochem. Biophys. Res. Commun.*, **361**, 1048–1053.

94 Stracke, R., Werber, M., and Weisshaar, B. (2001) *Curr. Opin. Plant Biol.*, **4**, 447–456.

95 Conrath, U., Beckers, G.J.M., Flors, V., García-Agustín, P., Jakab, G., Mauch, F., Newman, M.-A., Pieterse, C.M.J., Poinssot, B., Pozo, M.J., Pugin, A., Schaffrath, U., Ton, J., Wendehenne, D., Zimmerli, L., and Mauch-Mani, B. (2006) *Mol. Plant–Microbe Interact.*, **19**, 1062–1071.

96 Takabatake, R., Seo, S., Ito, N., Gotoh, Y., Mitsuhara, I., and Ohashi, Y. (2006) *Plant J.*, **47**, 249–257.

97 Leon, J., Rojo, E., and Sanchez-Serrano, J.J. (2001) *J. Exp. Bot.*, **354**, 1–9.

98 Garcês, H., Durzan, D., and Pedroso, M.C. (2001) *Ann. Bot.*, **87**, 567–574.

99 Arasimowicz, M., Floryszak-Wieczorek, J., Milczarek, G., and Jelonek, T. (2008) *Plant Biol.* doi: 10.1111/j.1438-8677.2008.00164.x.

100 Witte, M.B. and Barbul, A. (2002) *Am. J. Surg.*, **183**, 406–412.

101 Albina, J.E., Mills, C.D., Henry, W.L., Jr., and Caldwell, M.D. (1990) *J. Immunol.*, **144**, 3877–3880.

102 Albina, J.E., Abate, J.A., and Mastrofrancesco, B. (1993) *J. Surg. Res.*, **55**, 97–102.

103 Selamnia, M., Robert, V., Mayeur, C., Duée, P.H., and Blachier, F. (1998) *Biochim. Biophys. Acta*, **1379**, 151–160.

104 Wasternack, C., Stenzel, I., Hause, B., Hause, G., Kutter, C., Maucher, H., *et al.* (2006) The wound response in tomato – role of jasmonic acid. *J. Plant Physiol.*, **163**, 297–306.

105 Brady, J.D. and Fry, S.C. (1997) *Plant Physiol.*, **11**, 87–92.

106 Paris, R., Lamattina, L., and Casalongue, C.A. (2007) *Plant Physiol. Biochem.*, **45**, 80–86.

107 Wang, G., Moniri, N.H., Ozawa, K., Stamler, J.S., and Daaka, Y. (2006) *Proc. Natl. Acad. Sci. USA*, **103**, 1295–1300.

108 Kers, J.A., Wach, M.J., Krasnoff, S.B., Widom, J., Cameron, K.D., Bukhalid, R.A., Gibson, D.M., Crane, B.R., and Loria, R. (2004) *Nature*, **429**, 79–82.

109 Wach, M.J., Kers, J.A., Krasnoff, S.B., Loria, R., and Gibson, D.M. (2005) *Nitric Oxide*, **12**, 46–53.

110 Schimana, J., Gebhardt, K., Holtzel, A., Schmidt, D.G., Sussmuth, R., Muller, J., Pukall, R., and Fiedler, H.P. (2002) *J. Antibiot.*, **55**, 565–570.

111 Gusarov, I. and Nudler, E. (2005) *Proc. Natl. Acad. Sci. USA*, **102**, 13855–13860.

112 Wang, J. and Higgins, V.J. (2005) *Fungal Gen. Biol.*, **42**, 284–292.

113 Valderrama, R., Corpas, F.J., Carreras, A., Fernandez-Ocana, A., Chaki, M., Luque, F., Gomez-Rodriguez, M.V., Colmenero-Varera, P., Rio, L.A., and Barroso, J.B. (2007) *FEBS Lett.*, **581**, 453–461.

114 Ischiropoulos, H. (2003) *Biochem. Biophys. Res. Commun.*, **305**, 776–783.

11
Nitric Oxide Signaling by Plant-Associated Bacteria

Michael F. Cohen, Lorenzo Lamattina, and Hideo Yamasaki

Summary

The primary source of NO in the biosphere is energy metabolism within nitrifying and denitrifying bacteria, which are common inhabitants of the rhizosphere. NO produced by these bacteria regulates their mobility and gene expression and can elicit responses in neighboring organisms, including plants. Many other bacteria employ bacterial nitric oxide synthase to generate NO. One clear function for bNOS is in oxidative defense signaling. An endophytic association of actinomycetes and other bNOS-containing bacteria, some of which may be vertically transmitted within seeds, has been demonstrated among diverse plant species. Actinomycetes possessing bNOS activity display increased capacity for rhizosphere colonization, and two species have been shown to upregulate bNOS activity in response to plant-derived disaccharides. As yet, evidence of an effect of bNOS-derived NO on the plant host has been indirect. In contrast, there is strong evidence that denitrification-derived NO from *Azospirillum brasilense* can stimulate root branching in tomato, and this path of NO production is also responsive to plant host compounds. Further investigations will likely reveal other influences of bacterial NO in known plant NO-responsive pathways.

11.1
Introduction

The study of nitric oxide has entered the sphere of physiological ecology. Our understanding of NO roles in intercellular signaling within organisms has logically led to investigations of its capacity to mediate interactions between organisms. Surfaces and interiors of plants and other Eukarya are commonly colonized by bacteria, many of which can produce NO (Figure 11.1). In this chapter, we will focus primarily on findings that have come to light since a previous review of the field [1].

Nitric Oxide in Plant Physiology. Edited by S. Hayat, M. Mori, J. Pichtel, and A. Ahmad

ISBN: 978-3-527-32519-1

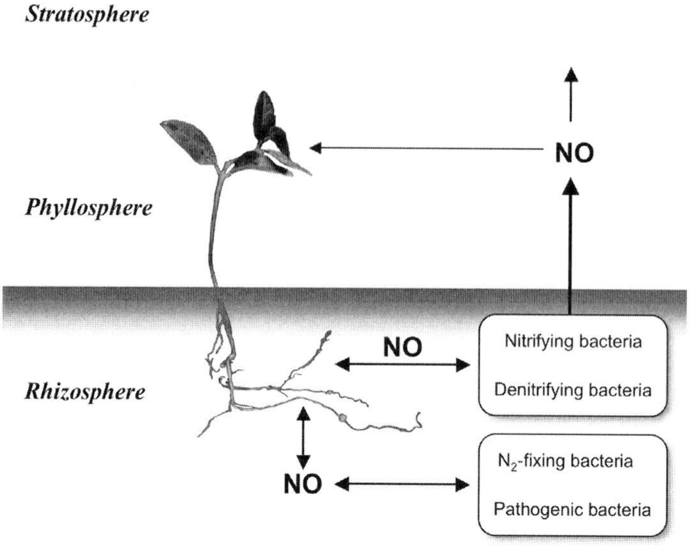

Figure 11.1 Major plant- and bacteria-related NO production sites and destinations. NO that escapes to the stratosphere is eventually oxidized and returns to earth as nitric acid (HNO_3).

11.2
Production of Nitric Oxide by Bacteria

Most biogenic NO is released as a product of bacterial energy metabolism. Some bacteria also produce NO specifically for cell signaling and, in one group, for nitration of an important secondary metabolite. At a practical level, the near ubiquity of NO-producing bacteria on and within plants, including inside seeds, must be considered in studies of presumptive plant NO synthesis; care should be taken beyond simple surface disinfection of plant surfaces to establish bacteria-free conditions [1].

The section headings below are organized by mode of metabolism and not as bacterial classifications; many bacteria posses metabolic flexibility that resists facile categorization. For instance, *Nitrobacter vulgaris* can nitrify under aerobic conditions but can switch to denitrifying metabolism under anaerobic conditions [2, 3].

11.2.1
Nitrification

Nitrification is the extraction of energy by the sequential oxidation of nitrogen that occurs as ammonia (Figure 11.2). This process is by far the dominant source of NO emissions from soils [1, 4–8]. Complete oxidation to nitrate is carried out by the combined action of two metabolically distinct bacterial groups. The first group, termed the nitrosofyers or ammonia-oxidizing bacteria, oxidizes ammonia to nitrite, while the second group, the nitrifying bacteria, oxidizes the nitrite to nitrate.

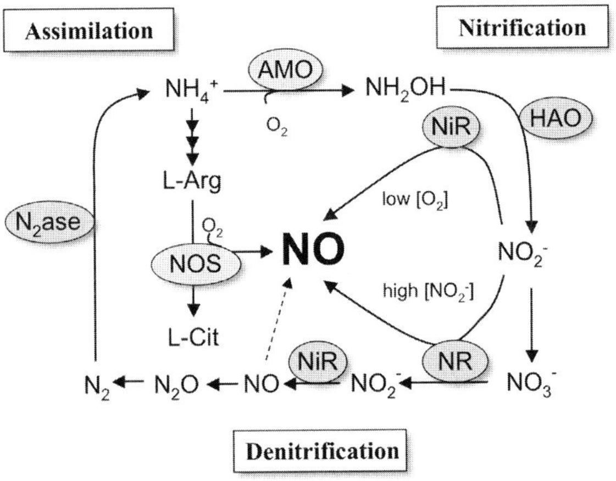

Figure 11.2 Pathways of bacterial NO production in relation to the nitrogen cycle. Selected enzymes shown in ovals are N_2ase, nitrogenase; AMO, ammonia monooxygenase; HAO, hydroxylamine oxidoreductase; NiR, nitrite reductase; NR, nitrate reductase; and NOS, nitric oxide synthase.

In the nitrosofyers, ammonia monooxygenase produces hydroxylamine, which is then oxidized to nitrite by hydroxylamine oxidoreductase. Maximal membrane potential is generated when the extracted electrons can flow to oxygen via a terminal cytochrome oxidase. However, under conditions where oxygen is limited, upregulation of nitrite oxidase activity consumes autogenic nitrite as a terminal electron acceptor, thereby releasing NO [9, 10].

11.2.2 Denitrification

In denitrification, nitrate is reduced in multiple steps to nitrogenous gases (Figure 11.2). This is the major mechanism whereby wetlands remove nitrogen from nitrate-rich treated municipal wastewaters [11, 12]. The complete denitrification process may take place in one species of microorganism or by successive action of multiple species. Many bacteria possess the first enzyme of the pathway, nitrate reductase (NR). In *Salmonella typhimurium*, when nitrate is not present, membrane-bound NR can reduce nitrite to NO [13], thus behaving similar to the NR of plants [14, 15]. Conversely, the NR of *Escherichia coli* does not directly catalyze NO formation; nitrate-cultivated *E. coli* cells produce NO from nitrite via a periplasmic nitrite reductase (NrfA) [16], which can also serve to detoxify NO to NH_4^+ [17, 18]. Such studies of NO flux in enteropathogens are likely to have relevance to plant-NO research since it is increasingly apparent that these bacteria are able to grow and persist in association with plants [19, 20].

In bacteria that possess a more complete denitrification pathway, which may release either N_2O or N_2 as the predominant product, NO is produced as a free intermediate by a periplasmic nitrite reductase [21, 22]. Concentrations of NO sufficient to cause toxicity and stimulate NO-defensive gene expression in neighboring bacteria can be released by denitrifying bacteria *en masse* [23].

Taking an evolutionary perspective, all mitochondrion-bearing eukaryotes share a direct tie with the lineage of denitrifying bacteria. From sequence homology studies, it is apparent that the electron transport system of mitochondria descended from that of a denitrifying bacterium [24, 25]. The capacity to reduce nitrite to NO in the mitochondria of animals, fungi [26], algae [27, 28], and plants [29] underscores this evolutionary connection with bacterial denitrifiers.

11.2.3 Nitric Oxide Synthase

Certain bacteria employ bacterial nitric oxide synthase (bNOS) to produce NO as a regulatory signal or, in at least one lineage, as a biosynthetic unit of a secondary metabolite. The NOS enzymes found in bacteria, Archea, and Eukarya domains all produce NO from L-arginine, releasing L-citruline as a by-product (Figure 11.2). Genetic confirmation of bNOS exists only for Gram-positive bacteria, including many common plant-associated species in the orders actinomycetales (*Mycobacterium, Nocardia, Rhodococcus,* and *Streptomyces*) and bacillales (*Bacillus, Geobacillus,* and *Paenibacillus*). The catalytic properties of bNOS are similar to the oxygenase domain of mammalian NOS but the bNOS lacks a reductase domain. In *Bacillus subtilis,* the reductase role may be taken by any of a variety of proteins; individual reductases capable of supporting sustained NO synthesis from bNOS *in vitro* [30] can be deleted without altering *in vivo* bNOS activity [31]. Caution should be taken in interpreting results of exposing live cells to traditional NOS inhibitors since some bacteria are capable of metabolizing these compounds, sometimes forming an NOS inducer in the process [32].

11.3 Regulatory Roles for Nitric Oxide in Bacteria

NO can transiently bind numerous functional sites of proteins, including heme, iron–sulfur clusters, and thiols, thus enabling NO to impact cell activities from the transcriptional to posttranslational levels [33]. Its short half-life and ability to diffuse across membranes make NO ideal for use as a near real-time signal between cells. Moreover, one-electron reduction of NO can potentially be mediated by various components of respiratory electron transport systems under oxygen-deficient conditions [34, 35], giving rise to HNO, which has bioactivity distinct from that of NO, and is the most reactive of the nitrogen oxides [36].

11.3.1 Metabolic Regulation

In denitrifying bacteria, NO promotes its own elimination by inducing transcriptional activators of enzymes that reduce NO to N_2O and inhibiting repressors (e.g.,

NsrR) of these genes [33]. Conversely, NO stimulates its own synthesis by inducing transcriptional activators of nitrite reductases. Thus, NO behaves as a signal to ensure coordinated respiratory flow of electrons through the denitrification pathway [33].

The influence of NO on gene expression in nitrifying bacteria is less well characterized [37]. Exposure of *Nitrosomonas europaea* to exogenous NO was found to induce NO production in this ammonia-oxidizing bacterium [38]. Consistent with this observation, acidification of a nitrite-containing medium, which results in nonenzymatic formation of NO [39], causes NsrR-dependent derepression of nitrite reductase (*nirK*) expression [40].

11.3.2
Regulation of Biofilm Formation

Nitric oxide has been shown to regulate biofilm formation in nitrifying and denitrifying bacteria. Biofilms confer several advantages to their resident bacteria, including increased resistance to antibiotics and a capacity for cooperative community metabolism [41]. In all of several species of ammonia-oxidizing species studied, NO was found to induce biofilm formation but at different concentrations [38]. Conversely, for the denitrifier *Pseudomonas aeruginosa*, NO caused biofilm dispersal and stimulated swarming motility [42].

Plant roots are common sites for biofilm formation [19]. Future research must seek to determine whether NO levels impact the dynamics of biofilm formation in the plant rhizosphere, which is commonly colonized by nitrifying and denitrifying bacteria [43, 44].

11.3.3
Stimulation of Oxidative and Nitrosative Defenses

Bacteria that produce NO or live near NO-emitting bacteria need to protect themselves against excessive NO levels and its decomposition products that can exert prooxidant effects. Under anaerobic conditions, NO can induce the expression of Nrf or a flavorubredoxin (FIRd), which reduces NO to ammonia or N_2O, respectively. When oxygen is present, expression of the genes that encode these detoxification enzymes is repressed, and instead cells will oxidize NO via a flavohemoglobin (Hmp) [33]. Expression of Hmp in the phytopathogen *Erwinia chrysanthemi* is necessary for virulence [45], as it protects the bacterium against nitrosative stress and limits the ability of the plant host to mount a hypersensitive response [46].

Microbes that inhabit the phytosphere are faced with a multitude of oxidative stresses. In the phyllosphere, they are exposed to reactive oxygen species (ROS) generated by exposure to UV light [47]. While in the rhizosphere, they must deal with high levels of ROS generated through the activity of plant oxidases [48] and protozoan feeding [49, 50]. One clear function for bNOS is in oxidative defense signaling. In the leaf endophyte *Rhodococcus* sp. strain APG1, bNOS activity is associated with elevated catalase activity and tolerance of the bacterium to H_2O_2 [51]. In *B. subtilis* and *Bacillus anthracis*, bNOS-derived NO exerts a cytoprotective effect by activating catalase and by inhibiting enzymatic reduction of free cysteine, thereby suppressing the production of hydroxyl radicals via the Fenton reaction [52, 53]. The fact that bNOS activity is necessary for engulfed *B. anthracis* cells to survive the oxidative burst in macrophages [52] may be of relevance to the observation that *B. anthracis* reproduces and exchanges DNA in the oxidative rhizosphere of plants [54]. *Streptomyces* spp. displaying NOS activity were found to be significantly more prevalent in the rhizosphere of apple root compared to surrounding soil [55]. For now, the idea that bNOS activity may confer protection from plant-derived oxidative stresses remains an enticing speculation that awaits direct experimental demonstration.

11.4
Bacterial Nitric Oxide in Plant–Bacteria Interactions

As described above, by acting as a signal to activate oxidative and nitrosative defense mechanisms, NO may improve the chances of bacterial survival in the phytosphere. In this section, we will describe examples of induction of bacterial NO synthesis in response to plants and known effects of bacterial NO on plant hosts.

11.4.1
Production of NO in Response to Plant Products

Plant-derived disaccharides have been shown to upregulate NOS activity in two actinomycetes, *Streptomyces turgidiscabies* and *Rhodococcus* APG1.

S. turgidiscabies, a pathogen of tubers, induces bNOS activity in response to the cell wall-building block cellobiose, which is released in infected tissues through the action of the phytotoxin thaxtomin. NO from bNOS nitrates the indole group of thaxtomin, a modification necessary for its toxicity [56, 57]. In the root infection process, NO is produced in excess of what is needed for nitration of thaxtomin [58], an observation that should lead to further investigation of potential effects of the diffusible NO on the physiology and development of the host plant. Many beneficial *Streptomyces* secrete cell wall-degrading enzymes in the process of colonizing a plant host, thereby releasing cellobiose into their immediate environment [59, 60]. It remains to be seen whether NO is produced in the course of these interactions as well.

Rhodococcus strain APG1, isolated as an endophyte of *Azolla pinnata*, induces bNOS activity in response to sucrose [51], which is of potential physiological significance since sucrose is exuded into the symbiotic leaf cavities of *Azolla* spp. [61].

11.4.2
Plant Responses to Bacterial NO: The *Azospirillum*–Tomato Interaction

The free-living diazotrophic *Azospirillum* is one of the best characterized plant growth promoting rhizobacteria (PGPR) [62]. *Azospirillum* can be considered a promiscuous plant colonizer since it is found in a wide range of habitats associated with roots of both monocots and dicots. *Azospirillum* fixes atmospheric N_2 through the nitrogenase complex under combined conditions of low nitrogen concentration and low O_2 tension [63, 64]. However, there is a lack of evidence supporting the role of nitrogen fixation by *Azospirillum* in facilitating plant growth processes; no significant increase in the nitrogen content of *Azospirillum*-inoculated compared to noninoculated crops has been observed [65].

Azospirillum is able to exert beneficial effects on plant growth parameters under certain environmental and soil conditions, for example, water status, nitrogen concentration and availability, temperature, and so on [66, 67]. An ability to modify root architecture has remarkably been a consistently observed aspect of *Azospirillum*-induced plant growth promotion. The most characteristic contributions of *Azospirillum* in promoting plant root growth are (i) an increased number of lateral and adventitious roots and (ii) positive effects on root hair formation and development. These changes in architecture and total surface area of roots enhance a plant's capacity to explore the soil. Significant differences in coleoptile growth pattern have also been observed between *Azospirillum*-inoculated and noninoculated plant species [68, 69].

The beneficial effects exerted by the association of *Azospirillum* with plant roots rely on molecular mechanisms that are still poorly understood. Many working hypotheses have been based on the ability of *Azospirillum* to synthesize phytohormones. The synthesis of auxins, cytokinins, and gibberellins has been the most commonly invoked reason to explain the growth promoting activity of inoculation with PGPR in general, not just *Azospirillum*. Indeed, auxins are thought to play a central role in the activity of beneficial rhizobacteria to induce significant changes in root architecture.

In parallel, during the past decade, many seminal studies on the biology of NO in plant physiology have appeared (see Ref. [70]). In plants, NO is a versatile signal molecule that controls and synchronizes a complex network of auxin-, cytokinin-, ethylene-, and abscisic acid-stimulated processes. As a result, NO profoundly affects plant growth, development, and stress physiology. Among the diversity of NO-mediated effects in plants, NO acts downstream of auxins to modulate adventitious and lateral root formation, as well as root hair development [71–73]. These NO-ascribed effects show a surprising and remarkable overlap with *Azospirillum*-induced effects on root architecture. Thus, we sought to determine whether there is any involvement of NO in the *Azospirllum*–plant root association. The first challenge was to determine the ability of *Azospirillum* itself to produce NO, then to investigate its origin and relevance in inducing plant responses to the bacterial inoculation.

Azospirillum produces NO under both aerobic and anaerobic conditions, using ammonium or nitrate as nitrogen source. More importantly, *Azospirllum* produces substantial amount of NO that is easily detectable through electron paramagnetic resonance (EPR) spectra of the NO–MGD–Fe adduct. *Azospirillum* produces a 6.4 nmol NO g^{-1} FW of bacteria growing in aerated OAB medium for 16 h [74]. This NO concentration is almost 10-fold higher than the NO concentration found in plants. When the NO was sequestered with the specific scavenger cPTIO, results clearly showed that the ability of *Azospirillum* inoculation to induce lateral root development in tomato was lost [74].

More recently, data from genetic approaches have given new insights that contribute to deciphering the function of *Azospirillum*-mediated NO production in its interaction with the plant root. First of all, it was demonstrated that *Azospirillum* generates more NO at the end of the log growth phase (stationary) than at the middle of log phase. Moreover, when nitrogen was supplied as nitrate, *Azospirillum* was able to generate NO by aerobic denitrification and produced 25-fold more NO than when nitrogen was supplied as ammonium. This NO concentration is quite high (more than 100 μM), and an endogenous NO-detoxifying system should be operational in *Azospirillum* when growing with nitrate. The periplasmic nitrate reductase isogenic mutant (Nap mutant *Faj164* [75]) of *Azospirillum brasilense* Sp245 was unable to generate high amounts of NO when growing with nitrate, confirming that a nitrate reductase in the periplasm is necessary for NO production in the presence of nitrate. Moreover, in contrast to the wild type, the *Faj164* mutant generated more NO when growing in ammonium than in nitrate [76]. In addition, the mutant *Faj009* is negative for indole-3-pyruvate decarboxylase activity and displays a 90% reduction in IAA synthesis [77] but is able to generate the same amount of NO as the wt strain Sp245 when growing in nitrate, confirming that the defect in auxin biosynthesis does not affect the activity of the periplasmic nitrate reducatse [76]. It was also confirmed that addition of the nitrification intermediate hydroxylamine enhances NO production in the presence of NH_4^+ as the nitrogen source, which is evidence of a heterotrophic nitrification pathway operating in *Azospirillum* [76]. These three isogenic strains, *A. brasilense* Sp245 wt, *Faj164* mutant, and *Faj009* mutant, are indeed powerful genetic tools for examining the ability of *Azosprillum* to modify root architecture in tomato and for gaining insight into the origin of NO in the association.

The wild-type strain of *A. brasilense* was compared with the two mutants *Faj009* and *Faj164* in their ability to induce adventitious and lateral roots in tomato. When the experiment was performed in water, the wild type and *Faj164* induced more lateral roots than the mutant impaired in IAA production, suggesting that auxins are required to attain some level of root branching [76]. The same results were reported in wheat [78, 79]. An interesting result was observed when the experiment was performed in the presence of nitrate. Under these conditions, both wild type and *Faj009* generated huge amounts of NO (25-fold more NO in nitrate than in ammonium). In contrast, mutant *Faj164*, impaired in the periplasmic NR, generated only 5% of NO compared to the wild type and *Faj009* in the presence of nitrate. *Faj164* was not able to induce adventitious and lateral root production even though this mutant synthesizes the same amount of IAA as the wild type. This result strongly supports the idea that NO produced by the bacteria might be critical for triggering the molecular mechanisms responsible for the metabolic and physiological changes that result in root branching [80].

The question of the number and relative importance of NO sources in the *Azospirillum*–plant root cell interaction is still open. Nitrite produced by Nap should serve as the substrate for NO production by the dissimilatory nitrite reductases found in *A. brasilense* [81]. Individual *A. brasilense* cells show induction of nitrite reductase expression on wheat root surfaces [81] and enhanced NO production upon exposure to seed extracts [82]. It is not yet known if *Azospirillum* inoculation triggers a mechanism that induces NO synthesis by root cells. Furthermore, it cannot be ruled out that NO production occurs in the apoplast of roots through a nonenzymatic reaction as was previously described in the aleurone cell layer of germinating barley seeds [83].

In conclusion, the finding that the level of aerobic NO synthesis in *Azospirillum* correlates with the level of lateral and adventitious root formation in tomato plants constitutes the first evidence establishing a plant response to bacterial-derived NO. Taken together, the results summarized above imply the existence of two-way signaling between *Azospirillum* and the plant host. Further studies should address the regulatory *cis*- and *trans*-acting elements that control the aerobic and anaerobic denitrification pathway and the modulator role that plant root components may play in an intimate association of *Azospirillum* with its partner.

11.4.3
Perspectives

Plants have evolved as open systems in habitats perfused with bacteria and the products of bacterial metabolism, including nitric oxide [84]. Thus, given our understanding of the roles of NO in various plant physiological and developmental pathways, it is almost certain that many influences of bacterial NO on plants remain to be discovered. Likewise, plants may release NO as a means to exert influence over their bacterial contingent.

As we endeavor to elucidate the pathways of NO production and responses in plants, an appreciation of the spatial and evolutionary relations between plants and bacteria can help to guide our investigations.

References

1 Cohen, M.F., Mazzola, M., and Yamasaki, H. (2006) Nitric oxide research in agriculture: Bridging the plant and bacterial realms, in *Abiotic Stress Tolerance in Plants: Toward the Improvement of Global Environment and Food* (eds A.K. Rai and T. Takabe), Springer-Verlag, New York, pp. 71–90.

2 Bock, E., Koops, H.-P., Möller, U.C., and Rudert, M. (1990) *Arch. Microbiol.*, **153**, 105–110.

3 Ahlers, B., König, W., and Bock, E. (1990) *FEMS Microbiol. Lett.*, **67**, 121–126.

4 Cohen, M.F. and Mazzola, M. (2006) *Plant Soil*, **286**, 1–12.

5 Dunfield, P.F. and Knowles, R. (1999) *Biol. Fertil. Soils*, **30**, 153–159.

6 Godde, M. and Conrad, R. (2000) *Biol. Fertil. Soils*, **32**, 120–128.

7 Jousset, S., Tabachow, R.M., and Peirce, J.J. (2001) *J. Environ. Eng.*, **127**, 322–328.

8 Venterea, R.T., Groffman, P.M., Verchot, L.V., Magill, A.H., Aber, J.D., and Steudler, P.A. (2003) *Global Change Biol.*, **9**, 346–357.

9 Jetten, M.S.M. (2001) *Plant Soil*, **230**, 9–19.

10 Whittaker, M., Bergmann, D., Arciero, D., and Hooper, A.B. (2000) *Biochim. Biophys. Acta*, **1459**, 346–355.

11 Bachand, P.A.M. and Horne, A.J. (1999) *Ecol. Eng.*, **14**, 17–32.

12 Xue, Y., Kovacic, D.A., David, M.B., Gentry, L.E., Mulvaney, R.L., and Lindau, C.W. (1999) *J. Environ. Qual.*, **28**, 263–269.

13 Gilberthorpe, N.J. and Poole, R.K. (2008) *J. Biol. Chem.*, **283**, 11146–11154.

14 Dean, J.V. and Harper, J.E. (1986) *Plant Physiol.*, **82**, 718–723.

15 Yamasaki, H., Sakihama, Y., and Takahashi, S. (1999) *Trends Plant. Sci.*, **4**, 128–129.

16 Corker, H. and Poole, R.K. (2003) *J. Biol. Chem.*, **278**, 31584–31592.

17 van Wonderen, J.H., Burlat, B., Richardson, D.J., Cheesman, M.R., and Butt, J.N. (2008) *J. Biol. Chem.*, **283**, 9587–9594.

18 Poock, S.R., Leach, E.R., Moir, J.W.B., Cole, J.A., and Richardson, D.J. (2002) *J. Biol. Chem.*, **277**, 23664–23669.

19 Rudrappa, T., Biedrzycki, M.L., and Bais, H.P. (2008) *FEMS Microbiol. Ecol.*, **64**, 153–166.

20 Brandl, M.T. (2006) *Annu. Rev. Phytopathol.*, **44**, 367–392.

21 Zumft, W. (1997) *Microbiol. Mol. Biol. Rev.*, **61**, 533–616.

22 Watmough, N.J., Butland, G., Cheesman, M.R., Moir, J.W.B., Richardson, D.J., and Spiro, S. (1999) *Biochim. Biophys. Acta*, **1411**, 456–474.

23 Choi, P.S., Naal, Z., Moore, C., Casado-Rivera, E., Abruna, H.D., Helmann, J.D., and Shapleigh, J.P. (2006) *Appl. Environ. Microbiol.*, **72**, 2200–2205.

24 de Vries, S. and Schröder, I. (2002) *Biochem. Soc. Trans.*, **30**, 662–667.

25 Castresana, C. and Saraste, M. (1995) *Trends Biol. Sci.*, **20**, 443–448.

26 Castello, P.R., David, P.S., McClure, T., Crook, Z., and Poyton, R.O. (2006) *Cell Metab.*, **3**, 277–287.

27 Tischner, R., Planchet, E., and Kaiser, W.M. (2004) *FEBS Lett.*, **576**, 151–155.

28 Bouchard, J.N. and Yamasaki, H. (2008) *Plant Cell Physiol.*, **49**, 641–652.

29 Planchet, E., Gupta, K.J., Sonoda, M., and Kaiser, W.M. (2005) *Plant J.*, **41**, 732–743.

30 Wang, Z.-Q., Lawson, R.J., Buddha, M.R., Wei, C.-C., Crane, B.R., Munro, A.W., and Stuehr, D.J. (2007) *J. Biol. Chem.*, **282**, 2196–2202.

31 Gusarov, I., Starodubtseva, M., Wang, Z.-Q., McQuade, L., Lippard, S.J., Stuehr, D.J., and Nudler, E. (2008) *J. Biol. Chem.*, **283**, 13140–13147.

32 Choi, W.S., Seo, D.W., Chang, M.S., Han, J.W., Hong, S.Y., Paik, W.K., and Lee, H.W. (1998) *Biochem. Biophys. Res. Commun.*, **246**, 431–435.

33 Spiro, S. (2007) *FEMS Microbiol. Rev.*, **31**, 193–211.

34 Garber, E., Wehrli, S., and Hollocher, T. (1983) *J. Biol. Chem.*, **258**, 3587–3591.

35 Poderoso, J.J., Lisdero, C., Schopfer, F., Riobo, N., Carreras, M.C., Cadenas, E., and Boveris, A. (1999) *J. Biol. Chem.*, **274**, 37709–37716.

36 Fukuto, J.M., Switzer, C.H., Miranda, K.M., and Wink, D.A. (2005) *Annu. Rev. Pharmacol. Toxicol.*, **45**, 335–355.

37 Klotz, M.G. and Stein, L.Y. (2008) *FEMS Microbiol. Lett.*, **278**, 146–156.

38 Schmidt, I., Steenbakkers, P.J.M., op den Camp, H.J.M., Schmidt, K., and Jetten, M.S.M. (2004) *J. Bacteriol.*, **186**, 2781–2788.

39 Yamasaki, H. (2000) *Philos. Trans. R. Soc. Lond. B*, **355**, 1477–1488.

40 Beaumont, H.J.E., Lens, S.I., Reijnders, W.N.M., Westerhoff, H.V., and van Spanning, R.J.M. (2004) *Mol. Microbiol.*, **54**, 148–158.

41 Danhorn, T. and Fuqua, C. (2007) *Annu. Rev. Microbiol.*, **61**, 401–422.

42 Barraud, N., Hassett, D.J., Hwang, S.-H., Rice, S.A., Kjelleberg, S., and Webb, J.S. (2006) *J. Bacteriol.*, **188**, 7344–7353.

43 Morgan, J.A., Martin, J.F., and Bouchard, V. (2008) *Wetlands*, **28**, 220–231.

44 Reddy, K.R., Patrick, W.H., and Lindau, C.W. (1989) *Limnol. Oceanogr.*, **34**, 1004–1013.

45 Favey, S., Labesse, G., Vouille, V., and Boccara, M. (1995) *Microbiology*, **141**, 863–871.

46 Boccara, M., Mills, C.E., Zeier, J., Anzi, C., Lamb, C., Poole, R.K., and Delledonne, M. (2005) *Plant J.*, **43**, 226–237.

47 Green, R. and Fluhr, R. (1995) *Plant Cell*, **7**, 203–212.

48 Cohen, M.F., Sakihama, Y., and Yamasaki, H. (2001) *Recent Res. Dev. Plant Physiol.*, **2**, 157–173.

49 Kreuzer, K., Adamczyk, J., Iijima, M., Wagner, M., Scheu, S., and Bonkowski, M. (2006) *Soil Biol. Biochem.*, **38**, 1665–1672.

50 Halablab, M.A., Bazin, M., Richards, L., and Pacy, J. (1990) *FEMS Microbiol. Immunol.*, **2**, 295–301.

51 Cohen, M.F. and Yamasaki, H. (2003) *Nitric Oxide*, **9**, 1–9.

52 Shatalin, K., Gusarov, I., Avetissova, E., Shatalina, Y., McQuade, L.E., Lippard, S.J., and Nudler, E. (2008) *Proc. Natl. Acad. Sci. USA*, **105**, 1009–1013.

53 Gusarov, I. and Nudler, E. (2005) *Proc. Natl. Acad. Sci. USA*, **102**, 13855–13860.

54 Saile, E. and Koehler, T.M. (2006) *Appl. Environ. Microbiol.*, **72**, 3168–3174.

55 Cohen, M.F., Yamasaki, H., and Mazzola, M. (2005) *Soil Biol. Biochem.*, **37**, 1215–1227.

56 Kers, J.A., Wach, M.J., Krasnoff, S.B., Widom, J., Cameron, K.D., Bukhalid, R.A., Gibson, D.M., Crane, B.R., and Loria, R. (2004) *Nature*, **429**, 79–82.

57 Wach, M.J., Kers, J.A., Krasnoff, S.B., Loria, R., and Gibson, D.M. (2005) *Nitric Oxide*, **12**, 46–53.

58 Johnson, E.G., Sparks, J.P., Dzikovski, B., Crane, B.R., Gibson, D.M., and Loria, R. (2008) *Chem. Biol.*, **15**, 43–50.

59 Suzuki, T., Shimizu, M., Akane, M., Hasegawa, S., Nishimura, T., and Kunoh, H. (2005) *Actinomycetologica*, **19**, 7–12.

60 Langlois, P., Bourassa, S., Poirier, G.G., and Beaulieu, C. (2003) *Appl. Environ. Microbiol.*, **69**, 1884–1889.

61 Peters, G.A. and Meeks, J.C. (1989) *Annu. Rev. Plant Physiol. Plant Mol. Biol.*, **40**, 193–210.

62 Bashan, Y. and Holguin, G. (1997) *Can. J. Microbiol.*, **43**, 103–121.

63 Döbereiner, J. and Day, J.M. (1976) Associative symbiosis in tropical grasses: characterization of microorganisms and dinitrogen fixing sites, in *Proceedings of the First International Symposium on Nitrogen Fixation* (eds W.E. Newton and C.J. Nyman), Washington State University Press, Pullman, WA, pp. 518–538.

64 Steenhoudt, O. and Vanderleyden, J. (2000) *FEMS Microbiol. Rev.*, **24**, 487–506.

65 Van de Broek, A., Dobbelaere, S., Vanderleyden, J., and Van Dommeles, A. (2000) *Azospirillum*–plant root interactions: signalling and metabolic interactions, in *Prokaryotic Nitrogen Fixation: A Model System for the Analysis of a Biological Process* (ed. E.W. Triplett), Horizon Scientific Press, Wymondham, pp. 761–777.

66 Bashan, Y., Holguin, G., and de Bashan, L.E. (2004) *Can. J. Microbiol.*, **50**, 521–577.

67 Okon, Y. and Labandera-Gonzalez, C.A. (1994) *Soil Biol. Biochem.*, **26**, 1591–1601.

68 Creus, C.M., Sueldo, R.J., and Barassi, C.A. (1998) *Can. J. Bot.*, **76**, 238–244.

69 Alvarez, M.I., Sueldo, R.J., and Barassi, C.A. (1996) *Cereal Res. Commun.*, **24**, 101–107.

70 Lamattina, L. and Polacco, J. (eds) (2007) *Nitric Oxide in Plant Growth, Development and Stress Physiology*, vol. 280, Springer-Verlag, Germany.

71 Lombardo, M.C., Graziano, M., Polacco, J.C., and Lamattina, L. (2006) *Plant Signal. Behav.*, **1**, 28–33.

72 Correa-Aragunde, N., Graziano, M., and Lamattina, L. (2004) *Planta*, **218**, 900–905.

73 Pagnussat, G.C., Simontacchi, M., Puntarulo, S., and Lamattina, L. (2002) *Plant Physiol.*, **129**, 954–956.

74 Creus, C.M., Graziano, M., Casanovas, E.M., Pereyra, M.A., Simontacchi, M., Puntarulo, S., Barassi, C.A., and Lamattina, L. (2005) *Planta*, **221**, 297–303.

75 Steenhoudt, O., Keijers, V., Okon, Y., and Vanderleyden, J. (2001) *Arch. Microbiol.*, **175**, 344–352.

76 Molina-Favero, C., Creus, C., Simontacchi, M., Puntarulo, S., and Lamattina, L. (2008) *Mol. Plant–Microbe Interact.*, **21**, 1001–1009.

77 Costacurta, A., Keijers, V., and Vanderleyden, J. (1994) *Mol. Gen. Genet.*, **243**, 463–472.

78 Dobbelaere, S., Croonenborghs, A., Thys, A., Vande Brooke, A., and Vanderleyden, J. (1999) *Plant Soil*, **212**, 155–164.

79 Barbieri, P. and Galli, E. (1993) *Res. Microbiol.*, **44**, 69–75.

80 Molina-Favero, C., Creus, C., Barassi, C., Lombardo, M.C., Correa-Aragunde, N., Lanteri, L., and Lamattina, L. (2008) Nitric oxide and plant growth promoting rhizobacteria: common features influencing root growth and development. in *Advances in Botanical Research*, vol. 46 (eds J.-C. Kader and M. Delseny), Academic Press, London, pp. 1–33.

81 Pothier, J.F., Prigent-Combaret, C., Haurat, J., Moenne-Loccoz, Y., and Wisniewski-Dye, F. (2008) *Mol. Plant–Microbe Interact.*, **21**, 831–842.

82 Pothier, J.F., Wisniewski-Dye, F., Weiss-Gayet, M., Moenne-Loccoz, Y., and Prigent-Combaret, C. (2007) *Microbiology*, **153**, 3608–3622.

83 Bethke, P.C., Badger, M.R., and Jones, R.L. (2004) *Plant Cell*, **16**, 332–341.

84 Yamasaki, H. (2005) *Plant Cell Environ.*, **28**, 78–84.

12
Nitric Oxide Synthase-Like Protein in Pea (*Pisum sativum* L.)

Mui-Yun Wong, Jengsheng Huang, Eric L. Davis, Serenella Sukno, and Yee-How Tan

Summary

Nitric oxide synthase activity was detected in pea (*Pisum sativum* L.) leaf extracts using a citrulline formation assay that is typically employed in mammalian systems. A total protein extraction method was modified from that used in mammalian systems based on biochemical activities such as the use of protease inhibitors, pH, and precipitation with salts and organic solvents. Physiological aspects in plants, such as effects of chemicals that induce systemic resistance to NOS activity and immunodetection of an NOS-like protein, were also studied. The NOS-like protein was partially isolated using liquid chromatography and characterized based on mammalian NOS inhibitor and cofactor requirements. Correlation of NOS activity and NOS-like gene expression during incompatible and compatible pea–bacteria interactions were investigated using interactions of *Ralstonia solanacearum* and *Pseudomonas syringae* pv. *pisi*, respectively, with pea. NOS activity was detected using citrulline formation assay. Gene expression was measured using real-time reverse transcription-polymerase chain reactions and a 348-bp probe designed from a cloned cDNA fragment of pea that was homologous to NOS of snail and AtNOS1/AtNOA1 of *Arabidopsis*. The possibility of NO production from various sources in cells of pea is also discussed.

12.1
Introduction

Nitric oxide (NO) has been recognized as an important signal molecule in various processes in both mammalian and plant systems. The source of NO in the mammalian system has been identified as the enzyme, nitric oxide synthase (NOS; EC1.14.13.39). NOS exists in various isoforms, namely, neuronal or brain NOS (nNOS or bNOS), inducible NOS in macrophages (iNOS), and endothelial NOS

Nitric Oxide in Plant Physiology. Edited by S. Hayat, M. Mori, J. Pichtel, and A. Ahmad

ISBN: 978-3-527-32519-1

(eNOS), named on the basis of the tissue from which they were originally extracted [1]. The NOS enzyme catalyzes the conversion of L-arginine to L-citrulline. However, the source of NO in plants has been intensely debated in recent years. Evidence has shown that NO in plants is produced via both enzymatic [2–8] and nonenzymatic reactions [9–11]. However, there is no clear evidence for the presence of a homologue to the mammalian NOS gene in plants, although arginine-dependent NOS activity has been documented [2, 3, 12–17].

Purifying an NOS protein or cloning an NOS gene from plants continues to be a challenge. In 2003, two groups of researchers, Chandok *et al.* [18] and Guo *et al.* [3], claimed to have isolated two plant proteins demonstrating NOS activity. These proteins were identified as a variant of the P protein (varP) of the glycine decarboxylase complex (GDC) in tobacco [18] and AtNOS1 in *Arabidopsis* that is similar to that in snail [3]. Simultaneously, our group had isolated a partial cDNA from pea with a sequence similar to *Arabidopsis* accession At3g47450 (unpublished data) that is the same sequence as *AtNOS1/AtNOA1*. However, Chandok *et al.* retracted their study results a year later and *AtNOS1* gene was later renamed as *AtNOA1* for nitric oxide-associated protein 1 [19] due to its indirect effect on NO synthesis. These phenomena demonstrate the complexity of NO synthesis in the plant system.

Multiple forms of NOS-like proteins may occur in plants that differ from their mammalian counterparts, upon which we will elaborate later on in this chapter. Since the process of NO synthesis in plants is not well understood, isolation of an NOS-like protein and cloning of the corresponding gene will greatly facilitate its understanding and will also help elucidate its role in plant defense against pathogen infection. In this chapter, we will describe our attempts to isolate and characterize an NOS-like protein and its corresponding gene in pea, and examine the relationship between NOS activity and expression of the related gene in pea bacterial disease development and resistance processes.

12.2 Physiological and Immunoblot Analyses of NOS-Like Protein of Pea

Nitric oxide is involved in varied biological processes in plants, including biotic interactions resulting in defense responses such as hypersensitive response (HR) and systemic acquired resistance (SAR) [15]. Plant defense mechanisms are associated with a rapid burst of oxidative reactions, changes in membrane ion fluxes, extracellular alkalinization, activation of signaling cascades, cell wall fortification, and phytoalexin and pathogenesis-related (PR) protein production [20]. The accumulation of NO and salicylic acid (SA) are essential in the induction of HR and SAR [21, 22]. Induced resistance responses such as SAR can be triggered by certain chemicals, nonpathogens, avirulent forms of pathogens, incompatible races of pathogens, or by virulent pathogens under circumstances where infection is stalled due to environmental conditions [23].

Pea (*Pisum sativum* L.), which previously demonstrated highest NOS activity compared to other plant species including soybean, tobacco, tomato, pepper, and corn (unpublished data), was studied in the current work. *Pseudomonas syringae* pv. *pisi*, which causes bacterial blight disease in pea, was selected for study on the basis of compatible interactions. *Ralstonia solanacearum*, a species taxonomically close to *P. syringae* and previously demonstrated to cause HR symptoms in infiltrated pea leaves [17], was used for incompatible interactions. Due to limited information on the involvement of NOS (and correspondingly NO) in plant interactions, we investigated the effects of abiotic and biotic interactions in pea on NOS activity. NOS activity in all preparations was determined by conversion of L-[U-^{14}C]arginine to L-[U-^{14}C]citrulline using the method of Bredt and Snyder [24] with modifications.

NO functions as a signal in plant disease resistance [15]. Copper chloride, Actiguard®, Triton-X100, and SA have been used to induce resistance in various species of plants in a systemic manner. Copper chloride induced phytoalexin production in rice via the jasmonic acid (JA) metabolic pathway [25]. Actiguard contains the active ingredient acibenzolar-s-methyl that functions through the same metabolic pathway as SA [26]. Triton-X100 and Tween 20 are surfactants used as wetting agents. In our work, the application of copper chloride, Actiguard, Triton-X100, and SA to pea leaves did not increase NOS activity significantly compared to the control [14], suggesting that NO functions upstream of the SA and JA signaling pathways of plant defense response. The results support previous observations by Wendehenne *et al.* [27]. The JA and SA pathways are two different signaling pathways activated after the induction of defense responses to insects and microbes, respectively. However, the pathways converge at a point controlled by NPR1 (a regulatory protein) and confer a similar enhanced defensive capacity against a broad spectrum of pathogens [23].

Information is limited on the involvement of NO in disease development compared to its involvement in disease resistance. NO functions as a signal in the induction of hypersensitive cell death, expression of early and late defense genes such as phenylalanine ammonia lyase (PAL) and pathogenesis-related (PR-1) protein, respectively, and phytoalexin accumulation [15, 16, 28]. HR is activated after the interaction of NO with reactive oxygen species (ROS), specifically H_2O_2 [21].

In our work, the temporal pattern of NOS activity in pea, that is, the occurrence of two peaks representing two NO bursts observed over time in both the incompatible and the compatible pea–bacteria interactions [14] suggests an inducible form of NOS, in contrast to a constitutive form in *Arabidopsis* [3]. The two peaks of NOS activity in both the incompatible and the compatible pea–bacteria interactions occurred at similar times suggesting that NOS is related not only to hypersensitive cell death in resistance response but also to necrotic cell death in disease development of the host. The occurrence of two NO bursts and the production of NO during a compatible interaction were also demonstrated by Delledonne *et al.* [15] and Conrath *et al.* [29]. NO is shown to be an important signal molecule involved in plant response to both biotic and abiotic stresses [12, 30–32].

Western blot analysis was performed to characterize the molecular size of pea NOS-like protein and to determine the reactivity of the protein with antibodies raised against mammalian NOS. The monomer molecular mass of mammalian NOS ranges from 130 to 160 kDa [21]. Huang and Knopp [17] reported the detection of a single immunoreactive band (approximately 55 kDa) in both extracts of the control and the leaves undergoing HR in tobacco after incubation with antibodies raised against mammalian NOS. The intensity of the immunoreactive band in HR tissues was much higher than that in control tissues. Ribeiro *et al.* [33] reported the detection of an approximately 166-kDa protein band after incubation with two different isoforms of mammalian anti-NOS antibodies in soluble fractions from young maize leaves and root tips. However, two bands of lower molecular weight were also detected, which were claimed to be degradation products of the 166-kDa protein.

Barroso *et al.* [34] reported the detection of an immunoreactive polypeptide of approximately 130 kDa in pea peroxisomal fractions using antibodies raised against mammalian iNOS. Our result is in contrast to that reported in Barroso's work where multiple immunoreactive protein bands were detected using three different isoforms of mammalian anti-NOS antibodies. No protein band of approximately 130 kDa in the crude protein extracts bound any antibody raised against mammalian NOS used; in contrast, however, a differential protein of approximately 30 kDa bound the mammalian anti-iNOS antibody in SDS-PAGE separations using protein extracts of HR tissues. Further analysis must be carried out to characterize this protein and determine its role in plant defense. The difference in the two studies could be due to the source of extracted protein and antibodies used. In the work of Barroso *et al.* [34], the use of samples containing only peroxisomal fractions for NOS detection would result in a much greater degree of iNOS antibodies binding specificity compared to the crude extracts that we used. However, the authors did not verify that the 130-kDa immunoreactive protein possessed NOS activity. Although both works used polyclonal antibodies for the immunodetection, they were from different sources and, thus, the degree of specificity of the antibodies could not be directly compared.

The detection of multiple immunoreactive bands in this study has two implications. First, the pea NOS-like protein may have a structure different from its animal counterpart, and thus, the size of the NOS-like protein and the ability to bind mammalian anti-NOS antibodies are not predictable. A similar conclusion was made by Barroso *et al.* [34] who reported that none of the NOS activities detected so far in plants [15, 33, 35] was found to be identical to the NOS isoforms present in mammals. To date, the only NOS-like protein in plants was isolated from *Arabidopsis* [3] that has sequence similarity to the protein implicated in NO synthesis in the snail *Helix pomatia* [36] and was related to NO production in hormonal responses in plants. The gene was later renamed as *AtNOA1* for nitric oxide-associated protein 1 [19] due to its indirect effect on NO synthesis. Second, using antibodies raised against mammalian NOS to detect a plant NOS-like protein must be interpreted cautiously and verified by functional assays of NOS activity in immunoreactive protein bands to counter false immunoreactivity. Butt *et al.* [37] used a proteomic approach to demonstrate that the mammalian NOS antibodies recognize many NOS-unrelated plant proteins and were of the opinion that it is inappropriate to infer the presence of plant NOS using immunological techniques.

Our findings, therefore, suggest that (a) NO functions upstream of SA in the signaling pathway of plant defense responses; (b) NO production is related to biotic stress in pea that leads to both resistance and disease development responses of the host; (c) antibodies raised against mammalian NOS did not have specificity in detecting an NOS-like protein in pea, suggesting that the pea NOS-like protein could be structurally different from mammalian NOS; and (d) immunodetection of a plant NOS-like protein must be conducted with caution and verified with functional assays.

12.3 Isolation and Characterization of an NOS-Like Protein of Pea

In mammalian cells, the production of NO is catalyzed by nitric oxide synthase that occurs in three isoforms named on the basis of the tissue source from which they were originally extracted: neuronal NOS (nNOS), also known as type I; inducible NOS in macrophages (iNOS), type II; and endothelial NOS, type III [1]. All NOS isoforms show 50–60% identity in their amino acid sequences [27]. Each NOS is a bidomain enzyme consisting of an N-terminal oxygenase and a C-terminal reductase [38]. The oxygenase domain contains a cytochrome P450 type heme center and a binding site for the cofactor tetrahydrobiopterin (BH_4). The reductase domain contains reduced nicotinamide adenine dinucleotide phosphate (NADPH), flavin adenine dinucleotide (FAD), and flavin mononucleotide (FMN) binding sites. Both oxygenase and reductase domains are connected by a calmodulin (CaM) binding site in the center of the enzyme. Each NOS has a different N-terminal extension determining the intracellular localization of the enzyme. In the active form, all NOS enzymes are homodimers.

The expression of nNOS and eNOS is constitutive whereas iNOS is inducible. The activity of constitutive NOS strictly depends on the elevation of intracellular free Ca^{2+} and requires the binding of CaM. Thus, both nNOS and eNOS show fast and transient activation (i.e., within minutes) [1, 39]. The activity of iNOS is independent of the intracellular free Ca^{2+} concentration and is sustained for a longer period, ranging from hours to days [1, 39].

A key step in the purification of the three isoforms of NOS includes 2′,5′-ADP (adenine 5′-diphosphate) affinity chromatography eluted with NADPH after crude sample preparation with ammonium sulfate precipitation or ion exchange chromatography [34, 40–42]. nNOS, eNOS, and iNOS have a molecular mass of approximately 160, 135, and 130 kDa, respectively. Both nNOS and eNOS require Ca^{2+}/CaM, NADPH, and BH_4 for their activity. On the contrary, iNOS requires NADPH, BH_4, FAD, and FMN but not Ca^{2+}/CaM for its activity. Other than mammalian cells, NOS proteins have also been isolated and purified from the bacterium *Nocardia* sp. [43], insects [44], snails [36], the fungus *Flammulina velutipes* [45], and the slime mold *Physarum polycephalum* [46] using various types of liquid chromatography such as ion exchange, affinity and gel filtration, and chromatofocusing. Although NOS activity has been documented in plants [15], the process of NO synthesis in plants is not well understood. Isolation of an NOS protein and/or cloning of the corresponding gene will greatly facilitate our understanding of NO synthesis and its role in plant defense against pathogen infection. However, these efforts have remained a challenge until recently, when one plant protein demonstrating NOS activity was identified, known as AtNOS1 [3]. The constitutively expressed AtNOS1 protein from *Arabidopsis thaliana* has sequence similarity to the protein implicated in NO synthesis in the snail *H. pomatia* [36]; it was also related to NO production in hormonal response in plants. The gene encoding AtNOS1, however, has no significant similarity to any gene encoding mammalian NOS.

Multiple forms of NOS-like protein may occur in plants, which differ from their counterparts in animal cells. Isolation of an NOS-like protein in pea will provide another source for comparison of the characteristics of NOS-like protein in plants. In addition, isolation of an NOS-like protein may be necessary to successfully clone the corresponding genes since the nucleotide sequence of NOS in plants has limited similarity to animal NOS. The objective of this study was to isolate and characterize a NOS-like protein in pea (*P. sativum* L.).

Our findings indicate that NOS activity was optimal when protein extractions were performed at pH 8.5 and 9.0 and at temperatures 23–25 °C [13]. These conditions differed from those in the fungus *F. velutipes* (pH 8.0 at 50 °C), the bacterium *Norcadia* spp. (pH 7.0–7.5 at 30 °C), and mammals (pH 7.5–7.8 at 37 °C) [41, 43, 45, 47]. These findings indicate that different species of NOS-like protein perform optimally under different conditions.

NOS protein extraction from various organisms typically includes a cocktail of protease inhibitors, metal ion chelators, reducing agents, and phenolic chelators [24, 43, 45, 46]. These chemicals are important components in the isolation process and they protect the protein of interest from proteolysis and oxidation [48]. In our work, adding 10 mM EGTA and 1 mM EDTA (metal ion chelator and metalloprotease inhibitor, respectively), 1 μM leupeptin and 1 mM PMSF (serine and thiol protease inhibitor, respectively), and 1% (w/v) PVPP (phenolic chelator) in the extraction buffer significantly retained NOS activity [13]. This was especially important in the isolation of NOS-like protein of pea since the protein rapidly lost its activity in each step of the isolation process. Using high doses of EGTA (10 mM) also facilitated the disruption and extraction of membrane-associated proteins [48].

Precipitation of crude protein extracts of pea with ammonium sulfate, sodium citrate, or sodium chloride at concentrations of 0.5–2.0 M caused substantial loss of NOS activity in the pellets after centrifugation compared to that in the supernatant of crude extract (i.e., control) [13]. However, after adding low molecular weight (LMW) filtrate (<8 kDa) to the pellets, precipitation with various concentrations of these salts increased NOS activity up to 77% compared to that of the control [13]. The precipitation of the NOS-like protein of pea with various concentrations of salts mentioned above was minimal or the salts caused the precipitated NOS-like protein to rapidly lose activity. The protein was not present in the supernatant because NOS activity was only 2.5–29% of that in the crude extract. On the contrary, NOS proteins of *P. polycephalum*, *F. velutipes*, and animals were successfully precipitated with 30–75% ammonium sulfate and with minimal loss in NOS activity [41, 45–47].

In contrast to precipitation with salts, precipitation of crude protein extracts with various concentrations of acetone and PEG resulted in NOS activity being detected mostly in the supernatant and minimally in the pellets [13]. Precipitation with 40% (v/v) acetone resulted in the highest increase in NOS activity (4.7-fold) in the supernatant compared to the control [13]. Substantial increase in NOS activity in the supernatant was possibly due to the release of NOS-like protein bound to membranes when a high dose of EGTA (10 mM) was used during protein extraction. Minimal activity in the pellets after precipitation with organic solvents (acetone and PEG) suggested that NOS-like protein of pea was membrane-bound. Proteins that are not precipitated by organic solvents are categorized as hydrophobic proteins, particularly those that are located in the cellular membranes [49]. The results of our work indicate that the NOS-like protein of pea might be located within cellular membranes; further work is needed to verify this hypothesis.

In mammals, nNOS is typically localized in neurons in both soluble and particulate forms and iNOS in macrophages in soluble form, while eNOS occurs in endothelial cells in particulate form [24, 41, 50]. An isoform of NOS localized in the inner membrane of rat liver mitochondria has also been reported [40]. Barroso *et al.* [34] demonstrated that a NOS-like protein of pea was localized in the matrix of peroxisomes (designated as perNOS) and chloroplasts using electron microscopic examination of immunogold labeling with antibodies against murine iNOS. Another NOS-like protein was detected either in the cytosol of cells in the division zone or nucleus of cells in the elongation zone of maize root tips (depending on the growth phase of cells) using antibodies against mouse macrophage NOS labeled with fluorescein isothiocyanate (FITC) [33].

In our work, the pea NOS-like protein was purified 16-fold and partial purification steps included treating crude extracts with 40% ethanol, passing the supernatant of the treated samples through an ion exchange DEAE-Sepharose column, and eluting bound proteins (including pea NOS-like protein) with a linear gradient of 0.135–0.2 M NaCl [13]. Treatment with acetone was substituted with ethanol due to the acetone sensitivity of Centricon Plus-20 cartridges used for ultrafiltration of treated samples. Treatment with 40% ethanol provided a comparable result in NOS activity as with 40% acetone. However, the protein did not bind to any of the affinity columns tested, that is, 2′,5′-ADP-Sepharose, ARG-Sepharose, or calmodulin-agarose as all NOS activity was detected in the flow-through. We had verified the working conditions of the 2′,5′-ADP-Sepharose column by applying the homogenate of sheep cerebellum to the column, following which a substantially high NOS activity was detected in the fraction eluted with 10 mM NADPH. Similarly, Lo *et al.* [51] found that homogenates of 3-day-old mung bean roots did not bind to 2′,5′-ADP-Sepharose or ARG-Sepharose. The fact that NOS-like protein of pea did not bind to 2′,5′-ADP-Sepharose and calmodulin-agarose might indicate that the protein lacked binding sites for cofactors NADPH and calmodulin. It is also surprising that pea NOS-like protein did not bind to ARG-Sepharose as all NOSs use arginine as substrate. Our findings suggest that pea NOS-like protein may be significantly different in structure from mammalian NOS.

Structural differences were also reported in an NOS-like protein identified in *Arabidopsis* (AtNOS1) [3]. The AtNOS1 protein was expressed in bacteria as a fusion protein with glutathione-*S*-transferase and purified 80-fold using glutathione-affinity chromatography. The 561 amino acid AtNOS1 protein has a molecular mass of 62 kDa, is constitutively expressed in the cells, and is involved in hormonal signaling. The cDNA sequence has homology to a sequence in the snail *H. pomatia* that encodes a protein implicated in NO synthesis and contains ATP/GTP binding domains [36]. However, the gene encoding AtNOS1 has no significant similarity to any gene encoding mammalian NOS and was later found to be indirectly involved in NO synthesis, thus, renamed as AtNOA1 for nitric oxide-associated protein 1 [19].

The challenge of isolating an NOS protein from plants is increasingly evident as shown in two studies. Although perNOS could be detected in the peroxisomes and chloroplasts of pea [34], to date neither its purification nor the identification of the gene encoding the protein has been reported. In addition, although an NOS gene has been identified in *Arabidopsis* by screening mutants [3], the protein itself was not purified from crude extracts. The isolation methods employed in animal systems were not helpful in isolating an NOS-like protein from plants, and alternative techniques need to be explored.

As a comparison to the characteristics of mammalian NOS and using citrulline formation assay, we demonstrated that the incubation of crude extracts of pea with known mammalian NOS inhibitors (aminoguanidine, N^G-monomethyl-L-arginine (L-NMMA), N^G-nitro-L-arginine methyl ester (L-NAME), and 7-nitroindazole) resulted in the inhibition of NOS activity in a concentration-dependent manner between 1 and 10 mM (unpublished data). At a concentration of 10 mM, NOS activity was inhibited 35% by aminoguanidine and L-NMMA, 65% by L-NAME, and 25% by 7-nitroindazole. In lupine roots and nodules, 56% of NOS activity was inhibited by 1 mM by L-NAME [35]. In maize leaves, NOS activity was reduced by 30% by a combination of 3 mM L-NAME and 3 mM aminoguanidine [33]. In pea peroxisome, NOS activity was inhibited completely by aminoguanidine and 90% by L-NAME at 1 mM [34]. NOS activity of AtNOS1 in *Arabidopsis* was inhibited by L-NAME [3]. In maize seedling leaves under dehydration stress, L-NAME inhibited NOS activity [12].

We have also shown that NOS activity based on citrulline formation of pea extracts was reduced 62 and 41%, respectively, in the absence of mammalian cofactors NADPH and Ca^{2+}/calmodulin (unpublished data). Other NOS cofactors such as flavin adenine dinucleotide, riboflavin 5′-phosphate (FMN), and tetrahydrobiopterin (BH_4) also reduced NOS activity, ranging from 15 to 33% (unpublished data). In lupine roots and maize tissues, NOS activity depended on Ca^{2+} [33, 35]. Pea peroxisomal NOS-like protein was also demonstrated to strictly depend on NADPH, and the absence of Ca^{2+} reduced more than 70% of NOS activity [34]. NOS activity of AtNOS1 in *Arabidopsis* depended on NADPH, calmodulin and Ca^{2+}, properties similar to mammalian eNOS and nNOS; however, the activity was not stimulated by BH_4, FAD, FMN, or heme [3].

Another challenge in isolating a pea NOS-like protein was the rapid loss of activity observed during the isolation process. This may be due to the disruption of protein configuration or the dissociation of some interacting proteins or unidentified cofactors. We found that adding the low molecular weight filtrate of pea leaf extracts (<8 kDa) to the proteins pelleted after precipitation by various concentrations of salts restored NOS activity at various degrees of significance [13]. The LMW filtrates may contain unidentified cofactors required for NOS activity that were dissociated from the pea NOS-like protein during the isolation process. Huang *et al.* [36] also reported that the activity of an NOS-like protein isolated from the snail (*H. pomatia*) might require interactions with other associated proteins.

In mammalian cells, NOS activity was affected by the interactions of the enzyme with more than 20 proteins [52]. For example, calmodulin binding was required for electron transfer [24]; NOSTRIN (NOS3 traffic inducer) binding was important for intracellular trafficking of NOS3 [53]; and Dynamin-2, Porin, and protein kinase B/Akt were required for activation [54–56]. The stability of NOS homodimers was also affected by protein–protein interactions. For example, the binding of PIN (a protein inhibitor of NOS1) to NOS1 or NAP110 (a protein inhibitor of NOS2) to NOS2 prevented dimerization of these enzymes, a configuration required for activation [57, 58].

In summary, the pea NOS-like protein was efficiently extracted from leaf tissues at pH 8.5 and 9.0. The addition of two protease inhibitors, EGTA and leupeptin, to the homogenization buffer significantly increased NOS activity at pH 8.5. Precipitation of the pea NOS-like protein with various concentrations of salts caused rapid loss of NOS activity, and the protein was not precipitated by organic solvents. The pea NOS-like protein was purified 16-fold and the partial purification steps included treatment of crude extracts with 40% ethanol, passing the supernatant of the treated sample through an ion exchange DEAE-Sepharose column and eluting bound proteins (including pea NOS-like protein) with a linear gradient of 0.135–0.2 M NaCl. The pea NOS-like protein did not bind to affinity columns (2′,5′-ADP-Sepharose, ARG-Sepharose, and calmodulin-agarose), suggesting that the protein lacks binding sites for NADPH, arginine, and calmodulin. NOS activity in pea leaf extracts was affected to varying degrees by known mammalian inhibitors and cofactors.

12.4
Molecular Cloning and Analyses of an NOS-Like Gene of Pea

In mammalian systems, cloning of cDNA encoding nitric oxide synthase of any isoform was performed by constructing a cDNA library from NOS-induced cells and identifying positive clones by plaque hybridization. The probe for plaque hybridization was prepared using PCR and primers designed based on either conserved binding sites for cofactors [59–61] or an amino acid sequence from isolated protein [62].

In *H. pomatia*, the cDNA encoding the 60-kDa NOS-like protein was isolated by probing a *Helix* expression cDNA library with an antibody to human neuronal NOS [36]. The cDNA sequence contained an open reading frame of 1377 bp (EMBL; GenBank accession no. X96994) and had a poly(A) tail but no stop codon. The cDNA sequence also contained an ATP/GTP binding site, and putative myristoylation sites, but did not contain consensus sequences for NADPH, FAD, ARG, and CaM binding sites.

In *P. polycephalum*, two similar cDNA sequences were isolated by screening a *Physarum* cDNA library with a 600-bp probe generated by PCR using primers designed to recognize either nucleotide sequences deduced from isolated peptides or consensus regions of other available NOS sequences from mammals, birds, insects, and a mollusk [46]. The two cDNAs were designated physnosa (3571 bp, GenBank accession no. AF145041) and physnosb (3316 bp, GenBank accession no. AF145040) and contained complete reading frames for NOS. Both sequences also contained binding motifs of mammalian NOS for FMN, FAD, NADPH, CaM, BH_4, zinc, and caveolin but lacked the binding site for calcium.

The cloning of NOS-like genes from plants has not been straightforward. The direct application of mammalian NOS sequences or binding motifs such as heme, FAD, and NADPH has not been fruitful for isolating and identifying NOS homologues in plants. In our work, the use of specific primers targeted at conserved functional domains of mammalian NOS such as heme, FAD, and NADPH has failed to clone an expressed NOS homologue from pea via reverse transcription-polymerase chain reaction (RT-PCR). This indicates the lack of similarity between plant NOS-like gene and mammalian NOS gene. Similar attempts using specific primers of *P. polycephalum* were also unsuccessful.

Subsequently, a database search of the *Arabidopsis* genome identified two potential plant NOS genes (Table 12.1): At4g09680 homologous to NOS in *Rattus norvegicus* (rat) and At3g47450 homologous to the br-1 protein associated with NOS in the snail *H. pomatia*.

Cloning of a homologue sequence of gene *At4g09680* in pea using the consensus sequence obtained from a multiple sequence alignment of rat nNOS, eNOS, and iNOS was unsuccessful even when cDNA of *Arabidopsis* was used as a control template in RT-PCR. The inability to amplify a cDNA fragment based on *Arabidopsis*

Table 12.1 Accessions At4g09680 and At3g47450 of *A. thaliana* ecotype Columbia obtained from TAIR and NCBI databases.

Name/locus	At4g09680	At3g47450
GenBank accession (cds)	NM_117036	NM_114613
GO cellular component	Chloroplast	Mitochondrion
Protein data		
Length (aa)	1075	569
Molecular weight (kDa)	118.945	62.806
Isoelectric point	6.3917	9.5636
Domains	—	ATP GTP A GTP-binding MMR HSR1
GenBank accession	CAB39635	CAB51217
Homology	NOS of *R. norvegicus* (rat)	NOS (br-1 protein) of *H. pomatia* (snail)

Table 12.2 NCBI nucleotide blast resulted in a perfect match of the 348-bp pea cDNA sequence and the corresponding coding region of gene *At3g47450* in *Arabidopsis* and AtNOS1/AtNOA1.

Sequences producing significant alignments	Score (bits)	*E* value
dbj|AK226195.1| *A. thaliana* mRNA for hypothetical pr...	641	0.0
ref|NM_180335.1|*A. thaliana* ATNOA1/ATNOS1/NOA1/NOS1...	641	0.0
ref|NM_114613.3|*A. thaliana* ATNOA1/ATNOS1/NOA1/NOS1...	641	0.0
gb|BT015887.1| *A. thaliana* At3g47450 gene, complete cds	641	0.0
gb|BT015353.1| *A. thaliana* At3g47450 gene, complete cds	641	0.0
emb|BX822723.1|CNS0A7TF *A. thaliana* full-length cDNA ...	641	0.0
emb|AL096860.1|ATT21L8 *A. thaliana* DNA chromosome 3, ...	220	3e−54

accession At4g09680 in either pea or *Arabidopsis* leaf tissues suggests that either the gene is hypothetical or it is not expressed in the tissues analyzed.

Subsequently, the attempt to clone a homologue sequence of gene *At3g47450* in pea resulted in the amplification of a 348-bp fragment. This gene appears to be present in the cells at very low copies since successful amplification required a high number of PCR cycles using cDNA of hypersensitive tissues rather than normal tissues. The 348-bp DNA sequence of pea is a perfect match with the corresponding coding region of gene *At3g47450* in *Arabidopsis* and AtNOS1/AtNOA1 (Table 12.2). Therefore, we named the *At3g47450* or AtNOS1/AtNOA1 homologue in pea, *PsNOA* gene. Efforts to clone and perform a functional analysis of the full-length *PsNOA* gene are in progress.

The At3g47450 accession was also determined by an independent laboratory to encode a protein associated with NOS activity in plant hormonal response that was named AtNOS1 [3]. The researchers were able to verify that AtNOS1 protein was expressed *in planta* and plays a vital role in plant growth, fertility, stomatal movements, and hormone signaling. *AtNOS1* cDNA was amplified using RT-PCR and confirmed by sequencing. The cDNA sequence has homology to a sequence in the snail *H. pomatia* that encodes a protein implicated in NO synthesis and contains ATP/GTP binding domains [36]. However, the gene encoding AtNOS1 has no significant similarity to any gene encoding mammalian NOS and was later found to be indirectly involved in NO synthesis and thus renamed as AtNOA1 for nitric oxide-associated protein 1 [19]. AtNOS1 orthologues were also found in rice and maize [63]. The localization of AtNOS1/AtNOA1 was shown to be targeted at mitochondria [64], similar to that of mouse AtNOS1 orthologue (mAtNOS1) [65]. Most recently, AtNOS1/AtNOA1 was found to be not an NOS but a member of the circularly permutated GTPase family (cGTPase), whose enzyme activity is necessary but not sufficient for its function *in planta* [66]. AtNOA1 was also shown to be colocalized with chloroplasts in leaves and is imported into isolated leaf chloroplasts [67].

Genomic DNA isolated from pea was subjected to Southern analysis to determine the copy number of the *PsNOA* gene. Digoxigenin (DIG)-labeled probe was successfully generated using the 348-bp fragment previously generated by RT-PCR and cDNA of pea. Pea genomic DNA digested with *Bam*HI and hybridized with the *PsNOA* DNA probe showed one band while digestion with *Hin*dIII produced at least four bands (data not shown). The results indicate that the *PsNOA* gene may consist of a small multigene family since multiple fragments could be amplified using the same probe.

In conclusion, the partial cDNA of *PsNOA* isolated from pea was found to be an orthologue of AtNOS1/AtNOA1 of *Arabidopsis*, rice, and maize [3, 63]. The role of *PsNOA* in NO production of pea needs to be determined.

12.5
Correlation Study of NOS-Like Gene Expression and NOS Activity in Compatible and Incompatible Pea–Bacteria Interactions

It is of interest to us to investigate the expression of the *PsNOA* gene during the interactions of pea with compatible bacteria, *P. syringae* pv. *pisi*, and incompatible bacteria, *R. solanacearum*, and relate gene expression to NOS activity in those interactions. Real-time RT-PCR and specific primers used to amplify the 348-bp *PsNOA* cDNA sequence mentioned in the above section were used in this study. NOS activity in all preparations was determined by conversion of L-[U-^{14}C]arginine to L-[U-^{14}C]citrulline using the method of Bredt and Snyder [24] with modifications.

In both *PsNOA* gene expression and NOS activity, two peaks of NO burst were observed (unpublished data). In the first peak, which occurred between 3 and 6 hpi (hour postinoculation of bacteria), *PsNOA* gene expression and NOS activity in both incompatible and compatible pea–bacteria interactions appeared to correlate. In the second peak, NOS activity was delayed to 24 hpi compared to the expression of *PsNOA* gene that occurred at 12 hpi in both interactions. Furthermore, in both cases, gene expression and NOS activity were generally highest in the compatible interactions, particularly at the two peaks, compared to those in incompatible interactions and the control. Our preliminary findings also indicate that *PsNOA* gene was expressed in an inducible manner in response to pathogen infection in both interactions resulting in resistance and disease development, in contrast to a constitutive form in *Arabidopsis* [3].

Our results on the two NO bursts concur with those of others. Conrath *et al.* [29] reported that in both the incompatible interaction between tobacco cells and avirulent *P. syringae* pv. tomato and the compatible interaction between soybean cells and virulent *P. syringae* pv. *glycinea*, two peaks representing two NO bursts were detected using membrane inlet mass spectrometry (MIMS) at approximately 1 h of treatment and followed by a prominent burst from 4 to 8 h [29]. However, the second NO burst in the compatible interaction was much less pronounced than that in the incompatible interaction. The authors concluded that the second NO burst is one likely key event in determining avirulence in plant–pathogen interactions. A similar finding was obtained by Delledonne *et al.* [15] using an oxyhemoglobin-/methemoglobin-based NO assay. Conrath *et al.* [29] used tissue culture cells for the interaction study whereas whole pea plants were used in our work, which may account for the different timing of NO burst. However, the *PsNOA* gene expression in leaf tissues was consistently highest during compatible interactions with pathogenic bacteria in both peaks in contrast to the results obtained by Conrath *et al.* [29] and Delledonne *et al.* [15] where NO detection was much lower in the second peak.

Data from gene expression study may be more reliable than that from NOS activity assay as the latter is influenced by many environmental factors including the availability of L-arginine in the tissues assayed, the efficiency of resin used to capture radiolabeled L-citrulline, and the sensitivity of the scintillation counter. Increases in L-arginine concentration resulted in an increase in the rate of L-citrulline formation and NO synthesis [35]. In our work, the concentration of L-arginine in crude extracts was not determined. Since there is no mammalian NOS homologue found in plants, NOS activity assay used in mammalian systems may not be appropriate for plants, possibly due to lack of functional domains that exist in mammalian NOS as mentioned earlier in this chapter [13, 68]. Alternative methods to detect arginine-dependent NOS activity such as electron paramagnetic resonance (EPR) spectroscopy and chemiluminesence were found to be reliable [69, 70].

The limited NOS activity observed during incompatible *R. solanacearum*–pea interactions was contrary to the hypothesis that a spike in NOS activity was primarily responsible for the generation of NO during HR. The difficulties in isolating an NOS-like protein in plants prompted early speculation that NO may be produced by many enzymes that are relevant to NO production such as nitrate reductase (NR), xanthine oxidoreductase, peroxidase, cytochrome P450, and some hemeproteins [71–75] similar to another important free radical, the superoxide anion ($O_2^{\bullet -}$) [30]. Furthermore, it has been proposed that NO emission in plants can be a generalized stress response to the accumulation of reactive oxygen species [76]. However, in recent times, two enzymes involved in mediating NO response in plants have received considerable attention, namely, NOS-like enzyme and NR [3, 8, 12–17, 69].

Our findings suggest that more than one NO-producing enzyme is involved in both the incompatible and the compatible pea–bacteria interactions, that is, PsNOA, a homologue of AtNOS1/AtNOA1, which was found to be a cGTPase, and an NOS-like enzyme (arginine-dependent enzyme). Confirmation of the NO-producing activity of the encoded PsNOA protein, together with corresponding knockout and complementation assays, will help establish its role in interactions with pathogenic bacteria. Furthermore, characterization of this new class of plant GTPases, particularly the uncharacterized protein domain, the CTD (C-terminal domain), is important to determine its function *in planta* as proposed by Moreau *et al.* [66].

References

1 Nathan, C. and Xie, Q.W. (1994) *Cell*, **79**, 915–918.

2 Corpas, F.J., Barroso, J.B., Carreras, A., Valderrama, R., Palma, J.M., Leon, A.M., Sandalio, L.M., and del Rio, L.A. (2006) *Planta*, **224**, 246–254.

3 Guo, F., Okamoto, M., and Crawford, N.M. (2003) *Science*, **302**, 100–103.

4 Rockel, P., Strube, F., Rockel, A., Wildt, J., and Kaiser, W.M. (2002) *J. Exp. Bot.*, **53**, 1–8.

5 Palma, J.M., Sandalio, L.M., Corpas, F.J., Romero-Puertas, M.C., McCarthy, I., and del Rio, L.A. (2002) *Plant Physiol. Biochem.*, **40**, 521–530.

6 Mansuy, D. and Boucher, J.L. (2002) *Drug Metab. Rev.*, **34**, 593–606.

7 Stohr, C., Strule, F., Marx, G., Ullrich, W.R., and Rockel, P. (2001) *Planta*, **212**, 835–841.

8 Yamasaki, H. and Sakihama, Y. (2000) *FEBS Lett.*, **468**, 89–92.

9 Stohr, C. and Ullrich, W.R. (2002) *J. Exp. Bot.*, **53**, 2293–2303.

10 Beligni, M.V., Fath, A., Bethke, P.C., Lamattina, L., and Jones, R.L. (2002) *Plant Physiol.*, **129**, 1649–1650.

11 Wojtaszek, P. (2000) *Phytochemistry*, **54**, 1–4.

12 Hao, G., Xing, Y., and Zhang, J. (2008) *J. Integr. Plant Biol.*, **50**, 435–442.

13 Wong, M., Huang, J., and Davis, E.L. (2007) *J. Biosci.*, **18**, 9–23.

14 Wong, M., Huang, J., and Davis, E.L. (2006) *J. Biosci.*, **17**, 87–97.

15 Delledonne, M., Xia, Y., Dixons, R.A., and Lamb, C. (1998) *Nature*, **394**, 585–588.

16 Durner, J., Wendehenne, D., and Klessig, F. (1998) *Proc. Natl. Acad. Sci. USA*, **9**, 10328–10333.

17 Huang, J. and Knopp, J.A. (1998) Involvement of nitric oxide in Ralstonia solanacearum-induced hypersensitive reaction in tobacco, in *Bacterial Wilt Disease: Molecular and Ecological Aspects* (eds P. Prior, J. Elphinstone, and C. Allen), INRA and Springer Editions, Berlin, pp. 218–224.

18 Chandok, M.R., Ytterberg, A.J., van Wijk, K.J., and Klessig, D.F. (2003) *Cell*, **113**, 469–482.

19 Crawford, N.M., Galli, M., Tischner, R., Heimer, Y.M., Okamobo, M., and Mack, A. (2006) *Trends Plant Sci.*, **11**, 526–527.

20 Hammond-Kosack, K.E. and Jones, J.D.G. (1996) *Plant Cell*, **8**, 1773–1791.

21 Delledonne, M., Zeier, J., Marocco, A., and Lamb, C. (2001) *Proc. Natl. Acad. Sci. USA*, **98**, 13454–13459.

22 Gaffney, T., Friedrich, L., Vernooij, B., Negrotto, D., Nye, G., Uknes, S., Ward, E., Kessmann, H., and Ryals, J. (1993) *Science*, **261**, 754–756.

23 van Loon, L.C., Bakker, P.A.H.M., and Pieterse, C.M.J. (1998) *Annu. Rev. Phytopathol.*, **36**, 453–483.

24 Bredt, D.S. and Snyder, S.H. (1990) *Proc. Natl. Acad. Sci. USA*, **87**, 682–685.

25 Rakwal, R., Tamogami, S., and Kodama, O. (1996) *Biosci. Biotechnol. Biochem.*, **60**, 1046–1048.

26 Okuno, T., Nakayama, M., Okajima, N., and Furusawa, I. (1991) *Ann. Phytopathol. Soc. Jpn.*, **57**, 203–211.

27 Wendehenne, D., Pugin, A., Klessig, D.F., and Durner, J. (2001) *Trends Plants Sci.*, **6**, 177–183.

28 Noritake, T., Kawakita, K., and Doke, N. (1996) *Plant Cell Physiol.*, **37**, 113–116.

29 Conrath, U., Amoroso, G., Kohle, H., and Sultemeyer, D.F. (2004) *Plant J.*, **38**, 1015–1022.

30 Bolwell, G.P. (1999) *Curr. Opin. Plant Biol.*, **2**, 287–294.

31 Tian, X. and Lei, Y. (2006) *Biol. Plant.*, **50**, 775–778.

32 Uchida, A., Jagendorf, A.T., Hibino, T., Takabe, T., and Takabe, T. (2002) *Plant Sci.*, **163**, 515–523.

33 Ribeiro, E.A., Cunha, F.Q., Tamashiro, W.M.S.C., and Martins, I.S. (1999) *FEBS Lett.*, **445**, 283–286.

34 Barroso, J.B., Corpas, F.J., Carreras, A., Sandalio, L.M., Valderrama, R., Palma, J.M., Lupianez, J.A., and del Rio, L.A. (1999) *Biol. Chem.*, **274**, 36729–36733.

35 Cueto, M., Hernandez-Perera, O., Martin, R., Bentura, M.L., Rodrigo, J., Lamas, S., and Golvano, M.P. (1996) *FEBS Lett.*, **398**, 159–164.

36 Huang, S., Kerschbaum, H.H., Engel, E., and Hermann, A. (1997) *J. Neurochem.*, **69**, 2516–2528.

37 Butt, Y.K., Lum, J.H., and Lo, S.C. (2003) *Planta*, **216**, 762–771.

38 Alderton, W.K., Cooper, C.E., and Knowles, R.G. (2001) *Biochem. J.*, **357**, 593–615.

39 Mayer, B. and Hemmens, B. (1997) *Trends Biochem. Sci.*, **22**, 477–481.

40 Tatoyan, A. and Giulivi, C. (1998) *J. Biol. Chem.*, **273**, 11044–11048.

41 Stuehr, D.J., Cho, H.J., Kwon, N.S., Weise, M.F., and Nathan, C.F. (1991) *Proc. Natl. Acad. Sci. USA*, **88**, 7773–7777.

42 Forstermann, U., Pollock, J.S., Schmidt, H.H., Heller, M., and Murad, F. (1991) *Proc. Natl. Acad. Sci. USA*, **88**, 1788–1792.

43 Chen, Y. and Rosazza, J.P.N. (1995) *J. Bacteriol.*, **177**, 5122–5128.

44 Regulski, M. and Tully, T. (1995) *Proc. Natl. Acad. Sci. USA*, **92**, 9072–9076.

45 Song, N.K., Jeong, C.S., and Choi, H.S. (2000) *Mycologia*, **92**, 1027–1032.

46 Golderer, G., Werner, E.R., Leitner, S., Grobner, P., and Werner-Felmayer, G. (2001) *Genes Dev.*, **15**, 1299–1309.

47 Hevel, J.M., White, K.A., and Marletta, M.A. (1991) *J. Biol. Chem.*, **266**, 22789–22791.

48 van Renswoude, J. and Kempf, C. (1984) *Methods Enzymol.*, **104**, 329–339.

49 Bollag, D.M., Rozycki, M.D., and Edelstein, S.J. (1996) *Protein Methods*, 2nd edn (eds D.M. Bollag, M.D. Rozycki, and S.J. Edelstein), Wiley-Liss, New York, pp. 7–24.

50 Pollock, J.S., Forstermann, U., Mitchell, J.A., Warner, T.D., Schmidt, H.H.H.W., Nakane, M., and Murad, F. (1991) *Proc. Natl. Acad. Sci. USA*, **88**, 10480–10484.

51 Lo, S.C.L., Cutt, Y.K., and Tam, M.F. (2000) *Nitric Oxide*, **4**, 207 (Abstract).

52 Nedvetsky, P.I., Sessa, W.C., and Schmidt, H.H.H.W. (2002) *Proc. Natl. Acad. Sci. USA*, **99**, 16510–16512.

53 Zimmermann, K., Opitz, N., Dedio, J., Renne, C., Muller-Esterl, W., and Oess, S. (2002) *Proc. Natl. Acad. Sci. USA*, **99**, 17167–17172.

54 Sun, J. and Liao, J.K. (2002) *Proc. Natl. Acad. Sci. USA*, **99**, 13108–13113.

55 Cao, S., Yao, J., McCabe, T.J., Yao, Q., Katusic, Z.S., Sessa, W.C., and Shah, V. (2001) *J. Biol. Chem.*, **276**, 14249–14256.

56 Dimmeler, S., Fleming, I., Fisslthaler, B., Hermann, C., Busse, R., and Zeiher, A.M. (1999) *Nature*, **399**, 601–605.

57 Ratovitski, E.A., Bao, C., Quick, R.A., McMillan, A., Kozlovsky, C., and Lowenstein, C.J. (1999) *J. Biol. Chem.*, **274**, 30250–30257.

58 Jaffrey, S.R. and Snyder, S.H. (1996) *Science*, **274**, 774–777.

59 Gnanapandithen, K., Chen, Z., Kau, C., Gorczynski, R.M., and Marsden, P.A. (1996) *Biochim. Biophys. Acta*, **1308**, 103–106.

60 Nakane, M., Schmidt, H.H.H.W., Pollock, J.S., Forstermann, U., and Murad, F. (1993) *FEBS Lett.*, **316**, 175–180.

61 Lyons, C.R., Orloff, G.J., and Cunningham, J.M. (1992) *J. Biol. Chem.*, **267**, 6370–6374.

62 Michel, T. and Lamas, S. (1992) *J. Cardiovasc. Pharmacol.*, **20**, S45–S49.

63 Zemojtel, T., Froblich, A., Palmicri, M.C., Kolanczyk, M., Mikula, I., and Wyrwicz, L.S. (2006) *Trends Plant Sci.*, **11**, 524–525.

64 Guo, F.Q. and Crawford, N.M. (2005) *Plant Cell*, **17**, 3436–3450.

65 Zemojtel, T., Kolanczyk, M., Kossler, N., Stricker, S., Lurz, R., Mikula, I., Duchniewicz, M., Scheilke, M., Ghafourifar, P., Martasek, P., Vingron, M., and Mundlos, S. (2006) *FEBS Lett.*, **580**, 455–462.

66 Moreau, M., Lee, G.I., Wang, Y., Grane, B.R., and Klessig, D.F. (2008) *J. Biol. Chem.*, **283**, 32957–32967.

67 Flores-Perez, U., Sauret-Gueto, S., Gas, E., Jarvis, P., and Rodriquez-Concepcion, M. (2008) *Plant Cell*, **20**, 1303–1315.

68 Neill, S., Bright, J., Desikan, R., Hancock, J., Harrison, J., and Wilson, I. (2008) *J. Exp. Bot.*, **59**, 25–35.

69 Corpas, F.J., Barroso, J.B., and Carreras, A. (2004) *Plant Physiol.*, **136**, 2722–2733.

70 Modolo, L.V., Augusto, O., Almeida, I.M.G., Magalhaes, J.R., and Salgado, I. (2005) *FEBS Lett.*, **579**, 3814–3820.

71 Rockel, P., Strube, F., Rockel, A., Wildt, J., and Kaiser, W.M. (2002) *J. Exp. Bot.*, **53**, 103–110.

72 Millar, T.M., Stevens, C.R., Benjamin, N., Eisenthal, R., Harrison, R., and Blake, D.R. (1998) *FEBS Lett.*, **427**, 225–228.

73 Huang, J., Sommers, E.M., Kim-Shapiro, D.B., and King, S.B. (2002) *J. Am. Chem. Soc.*, **124**, 3473–3480.

74 Boucher, J.L., Genet, A., Vadon, S., Delaforge, M., and Mansuy, D. (1992) *Biochem. Biophys. Res. Commun.*, **184**, 1158–1164.

75 Boucher, J.L., Genet, A., Valdon, S., Delaforge, M., Henry, Y., and Mansuy, D. (1992) *Biochem. Biophys. Res. Commun.*, **187**, 880–886.

76 Gould, K.S., Lamotte, O., Klinguer, A., Pugin, A., and Wendehenne, D. (2003) *Plant Cell Environ.*, **26**, 1851–1862.

13
Posttranslational Modifications of Proteins by Nitric Oxide: A New Tool of Metabolome Regulation

Jasmeet Kaur Abat and Renu Deswal

Summary

Posttranslational modifications of protein, whether reversible or irreversible, play important roles in cell signaling. PTMs regulate protein activity, intracellular localization, stability, and protein–protein interactions. Major PTMs include phosphorylation, glycosylation, methylation, and protein–sulfhydryl modifications. Nitric oxide, a key signaling molecule in plants, contributes to PTMs either in the form of cysteine nitrosylation, tyrosine and tryptophan nitration, thiol and tyrosine oxidation, or binding to metal centers. Of these, *S*-nitrosylation, the reversible attachment of NO to the free thiols of a protein, forming *S*-nitrosothiols, has been the most investigated. More than 100 *S*-nitrosylated proteins were identified from *Arabidopsis* and *Kalanchoe pinnata* including stress-related, metabolic, cytoskeletal, and signaling proteins. Target validation analysis has hypothesized functional relevance of *S*-nitrosylation in ethylene synthesis and signaling, cell death, photosynthesis, and biotic and abiotic stress conditions. Many targets still must be validated to appreciate their role(s). In addition, tyrosine nitration has been shown to play a role during salinity stress. Redox conditions in cells and organelles change temporally and spatially with abiotic and biotic signals. Global analysis of "*in vivo*" thiol status of proteins by techniques such as OxICAT in normal and stressed plants may assist in monitoring their physiological significance.

13.1
Introduction

Proteins, the product of gene expression, are primarily responsible for execution of the information contained within genes. Posttranslational modifications (PTMs) are chemical modifications of a protein that alter its properties either by proteolytic cleavage or by modification of a specific amino acid. PTMs regulate protein activity,

Nitric Oxide in Plant Physiology. Edited by S. Hayat, M. Mori, J. Pichtel, and A. Ahmad

ISBN: 978-3-527-32519-1

intracellular localization, stability, and protein–protein interactions. Major PTMs include phosphorylation, glycosylation, acetylation, methylation, ubiquitination, and protein–sulfhydryl modification [1]. These are both reversible and irreversible in nature, but of these, reversible modifications are more relevant to the regulation of metabolism [2]. These act as an on/off switch, controlling transmission of information. PTMs control plant metabolism by regulating cell signaling.

Nitric oxide (NO), a key signaling molecule, plays an important role in diverse physiological processes in plants [3–5]. Its synthesis from L-arginine is catalyzed by nitric oxide synthase (NOS), the presence of which in plants is still an enigma [6, 7]. NO is an unstable, diatomic gas and exists as a pollutant in the atmosphere. Being both water and lipid soluble, NO is able to diffuse through aqueous as well as lipid phases of the membrane; however, it has a very short half-life spanning only a few seconds [8].

NO can induce oxidative, nitrosative, and nitrative damage to DNA, lipids, and proteins [9]. It causes single and double strand breaks in DNA leading to mutations [10]. NO also nitrates membrane lipids and inhibits lipid peroxidation [11]. It is being established that PTMs of proteins by NO could be a prominent cellular signaling mechanism. NO contributes to PTMs either in the form of cysteine nitrosylation, tyrosine and tryptophan nitration, thiol and tyrosine oxidation, or binding to metal centers in proteins [12]. This chapter will mainly focus on recent developments in the field of *S*-nitrosylation and tyrosine nitration in plants.

13.2
S-Nitrosylation

S-Nitrosylation is the reversible attachment of NO to free thiols of a protein, forming *S*-nitrosothiols (RSNOs). The process is capable of modulating protein activity like a switch, similar to phosphorylation [13]. RSNOs are mainly formed by transnitrosation reactions (i.e., transfer of NO from *S*-nitrosoglutathione to free cysteine in the target protein). Reactive free thiols and disulfide bridges are important for tertiary structure of proteins. *S*-Nitrosylation induces disulfide bond formation in the neighboring thiols, thereby blocking free thiols and affecting protein activity. *S*-Nitrosylation can also result in the formation of sulfenic acid that in turn leads to the formation of sulfonic acid (SO_3^-), causing irreversible oxidative damage to proteins [14].

A single protein can contain more than one cysteine residue, but the basis of selection of the *S*-nitrosylation target residue is still not well understood. Earlier, the presence of an acid–base motif flanking the cysteine residue was considered essential [15]; however, other studies have suggested that it is not the primary structure but actually the three-dimensional environment that decides the sites of *S*-nitrosylation [16]. A recent study again emphasized the role of acid–base motifs [17], indicating that the issue regarding the selection of target cysteine still needs to be resolved.

S-Nitrosylation regulates a number of processes in animals and has been established as a redox-based signaling mechanism [15]. Recently, *S*-nitrosylated proteins were identified in *Arabidopsis* [18, 19] and *Kalanchoe pinnata* [20]. Diversity of the targets suggests their wide functional range. These include metabolic enzymes, abiotic and biotic stress-related proteins, cytoskeletal, photosynthesis, redox-related, and antioxidant enzymes. Of these, the majority are metabolic enzymes involved in carbon, nitrogen, and sulfur metabolism. Target validation has hypothesized functional relevance of *S*-nitrosylation in ethylene synthesis, cell death, photosynthesis, biotic, and abiotic stress conditions. In the following section, probable regulation of these processes by *S*-nitrosylation is discussed.

13.2.1 S-Nitrosylation and Ethylene Biosynthesis

Ethylene is a gaseous plant hormone influencing important plant processes such as germination and fruit ripening [21]. An overripe fruit promotes the ripening of other fruits in the vicinity through production of ethylene. NO and ethylene are both gaseous hormones, which act antagonistically. Earlier studies suggest that exogenous application of NO could delay fruit ripening (caused by ethylene). NO/ethylene stoichiometry was measured during strawberry *Fragaria anannasa* (Duch.) and avocado *Persea americana* (Mill.) ripening. It was observed that during fruit ripening, ethylene production increased while NO production decreased [22]. This finding suggested that exogenous application of NO could delay fruit ripening by inhibiting ethylene production, but the mechanism was not known. It was proposed that oxidative inactivation of ascorbate and Fe^{2+}, cofactors of the ethylene biosynthetic enzyme 1-aminocyclopropane 1-carbocylic acid (ACC) oxidase, was responsible for this inhibition [23].

Ethylene production is a multistep process regulated by various enzymes (Figure 13.1). Cobalamin-independent methionine synthase (Met synthase) and *S*-adenosyl methionine (SAM) synthase or methionine adenosyltransferase (MAT) are the enzymes of the methylmethionine cycle that provide SAM, the substrate for ethylene. SAM is acted upon by ACC synthase to generate ACC that is converted to ethylene by ACC oxidase. Met synthase and MAT were shown to be targets of *S*-nitrosylation in *Arabidopsis* [18].

Recently, it was shown that *S*-nitrosoglutathione (GSNO, 0.5 mM) treatment inhibits ethylene production in *Arabidopsis* cell culture. In the same study, it was shown that GSNO treatment inhibits MAT1 activity (one of the isoforms of MAT) by *S*-nitrosylation [24]. Using site-directed mutagenesis, Cys-114 was identified as the critical cysteine residue for MAT1. Another isoform of MAT was insensitive to *S*-nitrosylation.

The rate-limiting step of ethylene synthesis is the conversion of SAM to ACC that is catalyzed by ACC synthase [21]. Our work on identification of *S*-nitrosylated proteins from *Brassica juncea* has identified ACC synthase as a target of *S*-nitrosylation (unpublished work). Our data suggest that apart from indirectly controlling ethylene synthesis NO can directly control ethylene synthesis by regulating ACC synthase. These studies indicate a tight control of ethylene biosynthesis at multiple steps by NO.

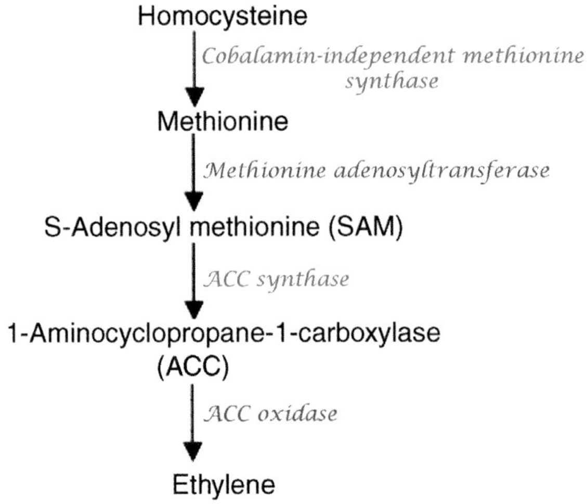

Figure 13.1 *S*-Nitrosylation targets in ethylene biosynthesis. Cobalamin-independent methionine synthase and methionine adenosyltransferase are the enzymes of the methylmethionine cycle that provide *S*-adenosyl methionine, a substrate for ethylene production. Enzymes depicted in green indicate *S*-nitrosylation targets reported only in plants; those in blue indicate *S*-nitrosylation targets reported both in animals and in plants; and those in black are not *S*-nitrosylated.

13.2.2
S-Nitrosylation and Photosynthesis

Photosynthesis is the most important metabolic process responsible for fulfilling the energy demands of the biosphere. Photosynthesis is a process involving the use of light energy to generate carbohydrates from CO_2 and water. NO induces stomatal closure [25], thus decreasing CO_2 availability and rate of photosynthesis. In mung bean, NO-responsive proteins (NORPs) were identified using a proteomics approach [26]. Sodium nitroprusside (SNP, 0.5 mM, NO donor) altered the protein expression of nine proteins, four of which were involved in photosynthesis. These included photosystem II oxygen evolving complex (OEC), sedoheptulose bisphosphate precursor, ribulose 1,5-bisphosphate carboxylase/oxygenase (Rubisco) subunit binding protein β-subunit, and Rubisco activase.

Photosynthesis in higher plants is composed of both light and dark reactions. Light reactions are responsible for oxygen generation, while the dark reactions involve carbon assimilation. Recent studies have shown that enzymes of both light and dark reactions are targets of NO.

Light reactions use two multisubunit photosystems (PSI and PSII) located in thylakoid membranes of the chloroplast. PSII is a multisubunit protein–pigment complex, including OEC and core protein complex. Members of both photosystems are targets of *S*-nitrosylation. PSII OEC is responsible for catalyzing the splitting of water to O_2 and $4H^+$. In *Arabidopsis*, PSII OEC protein 3 like [19],

PSII OEC 33 [18, 19], and oxygen evolving enhancer protein 3 precursor-like protein [19] were reported as targets of *S*-nitrosylation. PSII core protein complex comprises of five major integral membrane proteins D1, D2, 47 kDa, 43 kDa, and a cytochrome dimer. Along with 43-kDa protein, the 47-kDa protein performs the important function of linking the light harvesting complex of PSII to its reaction center. This important protein was also identified as the target of *S*-nitrosylation in *Arabidopsis* [18]. PSII D2 protein fragment is also a target of *S*-nitrosylation in *Arabidopsis* [18].

Dark reactions (carbon assimilation) are mediated via the Calvin cycle in the stroma of the chloroplast. In the Calvin cycle (Figure 13.2), carbon molecules from CO_2 are fixed into glucose. Various enzymes of the Calvin cycle including both large and small subunits of Rubisco [18–20], Rubisco activase [18], sedoheptulose bisphosphate (unpublished work), and transketolase ([18], unpublished work) are *S*-nitrosylated. Rubisco, the most abundant plant protein, plays an important role in photosynthesis. The cysteine residue was shown to be important for its catalytic efficiency [27]. Our studies have shown that *S*-nitrosylation inhibits Rubisco

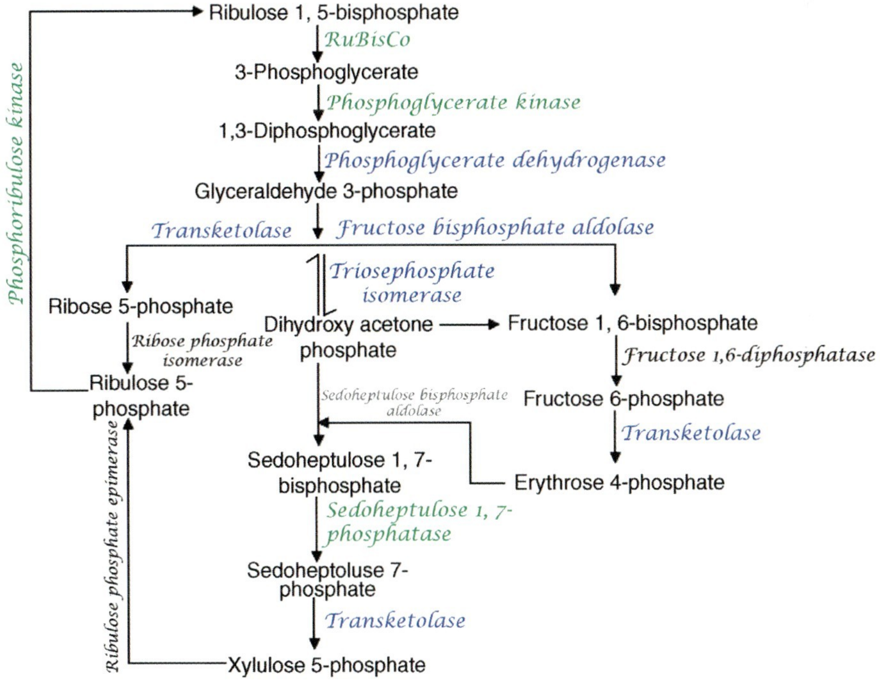

Figure 13.2 *S*-Nitrosylation targets in the Calvin cycle. During the Calvin cycle (dark reactions), carbon molecules from CO_2 are fixed into glucose. ATP and NADH generation or utilization is not depicted. Enzymes depicted in green indicate *S*-nitrosylation targets reported only in plants; those in blue indicate *S*-nitrosylation targets reported both in animals cytosolic forms and in plants; and those in black are not reported as *S*-nitrosylation targets.

carboxylase activity in a dose-dependent manner. This inhibition was observed both in crude and in purified Rubisco from *K. pinnata* [20]. Rubisco is the only validated *S*-nitrosylated target of photosynthesis to date. Similar results have also been obtained with purified Rubisco from *B. juncea* (unpublished work).

Sedoheptulose bisphosphate (SBPase; EC 3.1.3.37), which functions in the regenerative phase of the Calvin cycle, is also *S*-nitrosylated. It catalyzes the dephosphorylation of sedoheptulose 1,7-bisphosphate to sedoheptulose 7-phosphate (Figure 13.2). Its importance was shown in transgenic tobacco plants where a small reduction in SBPase activity resulted in the reduction of photosynthetic carbon fixation [28]. Transketolse (EC 2.2.1.1), another Calvin cycle enzyme, catalyzes the conversion of sedoheptulose 7-phosphate and glyceraldehyde 3-phosphate to pentoses (D-ribose 5-phosphate and D-xylulose 5-phosphate). Neither sedoheptulose bisphosphate nor transketolase are validated experimentally. Whether *S*-nitrosylation inhibits or activates these would become clear after validation.

13.2.3
S-Nitrosylation and Glycolysis

Glucose is the universal metabolic substrate and major fuel for most organisms. NO not only regulates its synthesis, as described above, but also its utilization. Glycolysis is the catabolic pathway responsible for breakdown and utilization of glucose. Glycolysis is a universal central pathway and differs among species only in its mode of regulation. Like photosynthesis, glycolysis is also a multistep process (Figure 13.3) where a molecule of glucose breaks down to form two molecules of pyruvate. In animals, upregulation of glycolysis by NO was reported. NO upregulated the breakdown of glucose by increasing the activity of 6-phosphofructo-1-kinase [29]. In plants also, NO donor (SNP) treatment resulted in glucose depletion in mung bean [26]. The mechanism for this glucose depletion is unknown, but two probable mechanisms were suggested, either inhibition of photosynthesis leading to less glucose production or its higher catabolism by upregulation of glycolysis.

Recent studies have elucidated several enzymes of glycolysis as targets of *S*-nitrosylation including fructose bisphosphate aldolase [18, 20], triose phosphate isomerase [18–20], glyceraldehyde 3-phosphate dehydrogenase (GAPDH) [18, 20], phosphoglycerate kinase [18–20], and enolase [18]. The presence of most of the glycolytic enzymes as targets of *S*-nitrosylation suggests that glycolysis might be tightly regulated at each step by NO via *S*-nitrosylation. Of all these targets, only GAPDH was experimentally validated. GSNO (100 μM) treatment to crude extracts of *Arabidopsis* cell cultures resulted in 90% reduction in GAPDH activity [18].

NO also regulates the enzymes of the glycolysis feeder pathway. Apart from glucose, other stored complex carbohydrates such as galactose, sucrose and trehalose, and so on are transformed into glycolytic intermediates via the feeder pathway. UDP-glucose-4-epimerase is an important enzyme of the feeder pathway required for conversion of D-galactose to UDP-glucose. The glucose thus formed enters the glycolytic pathway. This enzyme was reported as a target of *S*-nitrosylation in *K. pinnata* [20]. Whether *S*-nitrosylation activates or inhibits its activity is still unknown.

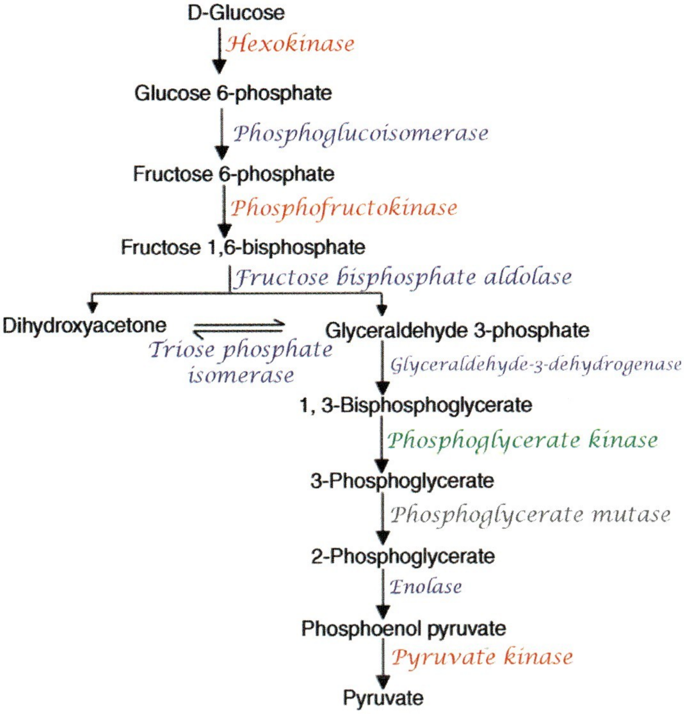

Figure 13.3 *S*-Nitrosylation targets in glycolysis. One molecule of glucose is converted to two molecules of pyruvate, releasing two molecules of ATP and NADH each. ATP and NADH generation is not depicted. Enzymes depicted in red indicate *S*-nitrosylation targets reported only in animals; those in green indicate *S*-nitrosylation targets reported only in plants; and those in blue indicate *S*-nitrosylation targets reported both in animals and in plants.

Evidence suggests that NO strongly influences carbohydrate metabolism including glucose synthesis, its catabolism, and even the intermediate feeder pathway. Thus, by regulating carbohydrate metabolism, NO might affect crop yields. Direct effects of these modifications on crop yield need to be analyzed in detail.

13.2.4
S-Nitrosylation and Biotic/Abiotic Stresses

Biotic and abiotic stresses significantly limit agricultural productivity. NO production occurs as a generalized stress response [30]. Several reports suggest involvement of NO during biotic stress response (i.e., plant–pathogen interactions). Inactivation of metacaspase 9 (plant caspases) of *Arabidopsis* by *S*-nitrosylation of critical cysteine residue was reported while the mature form was found insensitive to *S*-nitrosylation [31]. *S*-Nitrosoglutathione reductase (GSNOR) modulates cellular NO by using GSNO (NO donor) as the substrate and thus removing it from the cellular pool. In *Arabidopsis*, *S*-nitrosothiols were shown to regulate plant disease resistance [32]. Loss of AtGSNOR1 (*Arabidopsis* GSNOR) function led to an increase in *S*-nitrosothiol levels and a decrease in plant defense responses to pathogen attack. Our work on identification of *S*-nitrosylated proteins in *K. pinnata* identified a putative disease resistance protein that showed homology to cytoplasmic nucleotide binding site/leucine-rich repeat (NBS/LRR) proteins [20]. Recently, an increase in *S*-nitrosothiols was shown during progression of HR in *Arabidopsis* [19]. Expression of 18 *S*-nitrosylated proteins varied during defense response. These proteins belonged to different protein families such as metabolic enzymes, redox-related, stress-related, and signaling proteins [19].

Abiotic stresses not only limit agricultural productivity but also limit the geographical distribution of plants. Major abiotic stresses include salinity, drought, and low and high temperatures. The following section focuses on the regulation of stress relevant proteins by *S*-nitrosylation.

Salinity is one of the major abiotic stresses affecting plant productivity. It was shown that NaCl induced a transient increase in NO in maize leaves that in turn enhanced salt tolerance through increased activities of the proton pump and Na^+/H^+ antiport in the tonoplast [33]. In olive seedlings, it was demonstrated that salt stress increased L-arginine-dependent NO production, total SNO, and number of proteins that underwent tyrosine nitration, suggesting generation of reactive nitrogen species (RNS) [34]. The role of NO in imparting drought tolerance was shown in wheat where treatment with NO donor induced stomatal closure ultimately resulting in drought tolerance [35].

Abscisic acid (ABA, plant hormone) confers drought tolerance. NO was shown to contribute to ABA-induced stomatal closure. NO regulates K^+ and Cl^- channels by activating ryanodine-sensitive Ca^{2+} channels in guard cells of *Vicia* [36]. In a recent study, the same group showed that *S*-nitrosylation of the K^+ channel or an associated regulatory protein can regulate stomatal movements [37].

Low and high temperatures also affect plant growth and hence agricultural productivity. It was reported that in tomato, wheat, and corn application of NO donor mediated chilling tolerance, probably by suppressing high levels of reactive oxygen species (ROS) that accumulate during stress [3]. In *Arabidopsis*, GSNOR was also shown to play an important role in thermotolerance [38]. Arabidopsis thermotolerance defective mutant hot 5 was identified and the gene HOT 5 was determined to be encoding GSNOR. Using hypocotyl elongation and thermotolerance assay with hot 5 mutants, it was shown that the loss of GSNOR function retards growth and inhibits thermotolerance. Recent studies with *B. juncea* indicated that during low-temperature (LT, 4 °C) conditions, total *S*-nitrosothiols increased 1.4-fold. Biotin switch assay [39] coupled with 2D-gel electrophoresis showed LT-induced changes in *S*-nitrosoproteome of *B. juncea*. A total of 67 *S*-nitrosylated proteins were detected in the pH range 3–10. Of these, 17 spots showed intensity variation with 9 spots showing an increase while 8 showed a decrease in *S*-nitrosylation in response to LT. Identification of these spots is underway (unpublished work).

Involvement of *S*-nitrosylation in multiple pathways through regulation of enzymes of these pathways demonstrates its importance as a PTM. In animals, *S*-nitrosylation has been established as an important redox-based modification, mediating the effects of NO. In plants, studies are focused on revealing as many targets of *S*-nitrosylation as possible. Validation of these targets would be required to understand the physiological significance of *S*-nitrosylation.

Another PTM mediated by NO is tyrosine nitration. In contrast to *S*-nitrosylation, it can cause irreversible modification of proteins leading to the loss of function. Details about it are presented in the following section.

13.3 Tyrosine Nitration

Tyrosine nitration is a covalent protein modification, resulting in the incorporation of a nitrotriatomic group ($-NO_2$), at position 3 of the phenolic ring of a tyrosine residue, forming 3-nitro-tyrosine [40]. Tyrosine nitration could be either heme protein (metalloprotein) catalyzed or nonenzymatic (peroxynitrite mediated) [12]. Some metalloproteins such as myoglobin, cytochrome P450, and others catalyze the oxidation of nitrite to NO, resulting in tyrosine nitration. NO also reacts with superoxide (O_2^-) to form peroxynitrite ($ONOO^-$). Peroxynitrite is a strong oxidant leading to cellular injury. Like *S*-nitrosylation, tyrosine nitration can also alter protein function, conformation, and most importantly impose steric restrictions thereby inhibiting tyrosine phosphorylation [41]. Selectivity of a protein as a target of tyrosine nitration depends upon many factors, for example, localization of protein, number of tyrosine residues, and the presence of a specific amino acid sequence that promotes nitration [42].

In animals, tyrosine nitration is clinically used as a marker for disease onset and progression. Several studies indicate that during onset of disease, reactive oxygen species increase, which react with NO to produce $ONOO^-$. A number of proteins have been shown to undergo tyrosine nitration including cytochrome *c*, protein kinase Cε, fibrinogen, calmodulin, and actin [43]. Apart from proteins, $ONOO^-$ also reacts with other biomolecules such as nucleic acids and lipids.

Compared to animals, information regarding tyrosine nitration in plants is lacking. Recently, a few studies have shown the importance of tyrosine nitration in plant development and stress tolerance. Using $ONOO^-$-specific fluorescent probe (aminophenyl fluorescene, AFP) $ONOO^-$ generation was detected in tobacco BY-2 cells. INF-1 (elicitin) treatment induced $ONOO^-$ generation within 1 h, which reached a maximum level at 6 h [44]. INF-1 induced $ONOO^-$ resulted in tyrosine nitration of two polypeptides (20 and 50 kDa). Expression of the 20-kDa band was suppressed by urate ($ONOO^-$ scavenger) while that of 50 kDa was constant. These data suggest that $ONOO^-$ generation/tyrosine nitration occurs during defense response in plants.

In olive seedlings, salt stress was also shown to increase the number of proteins undergoing tyrosine nitration [34]. Tyrosine nitration was detected using antinitrotyrosine antibodies, which are commonly used in animal systems for detection of tyrosine nitration. Immunolocalization experiments using antinitrotyrosine antibodies suggested that vascular tissues could play a role in redistribution of reactive nitrogen species.

As both biotic and abiotic stresses increase tyrosine nitration of proteins, this process can possibly be used as a stress marker as it is used as a disease marker in animals.

Besides the above two prominently analyzed PTMs, NO also acts by binding to metal centers as described below.

13.4 Binding to Metal Centers

NO binds to metal centers in metalloproteins. Soluble guanylate cyclase (sGC, EC 4.6.1.2) is the major heme-containing protein to which it binds. Binding of NO to iron in the heme moiety of sGC induces a conformational change, resulting in the formation of cyclic guanosine monophosphate (cGMP) from guanosine triphosphate (GTP) [45]. This binding is reversible and sGC turns off when NO levels are reduced. cGMP thus formed acts as a second messenger controlling a vast number of signaling cascades in the cell. cGMP-dependent NO signaling is common in animals.

Many cGMP-mediated processes are known in plants, but only recently a plant homologue of sGC (AtGC1) was cloned from *Arabidopsis* [46]. Overexpression of AtGC1 in *E. coli* induced 2.5 times more cGMP than in the control. Classical GCs are characterized by the presence of a heme (NO) binding, catalytic, transmembrane, dimerization, kinase like, ligand binding, signal peptide, and functionally unassigned domain. AtGC1 is quite distinct from other known sGCs in animal and bacterial systems. AtGC1 has an unusual reduced N-terminal catalytic domain compared to other GCs; in addition, it does not share any significant sequence similarity with the rest of the domains, suggesting it to be a member of a new class of GC.

13.5 Conclusions and Prospects

It is clearly established that a simple molecule such as NO regulates a broad spectrum of processes in animals and plants. At present, our focus is on the identification of NO targets in the form of either *S*-nitrosylated or tyrosine-nitrated proteins (Figure 13.4). In future, the focus would be to validate these targets and know how *S*-nitrosylation regulates their activity and which are the critical cysteines targeted by *S*-nitrosylation. The thiol status of a protein is very important for its catalytic activity [47]. *S*-Nitrosylation is the PTM that blocks the free thiols. Analysis of thiol status of proteins in normal and stressed plants, using techniques such as OxICAT (a combination of ICAT – isotope-coded-affinity tag – and differential thiol trapping technique) [48], could complement our data and provide functional significance. This technique is most helpful as it not only allows quantification of thiol modification but also identifies the protein and the target cysteine(s). It is important to identify

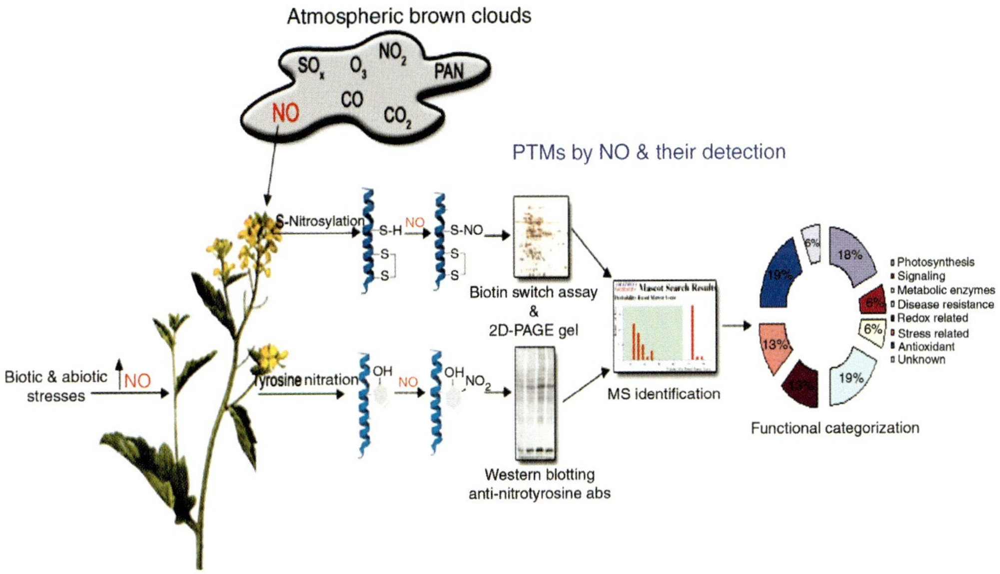

Figure 13.4 *S*-Nitrosylation and tyrosine nitration of proteins by NO detection and identification. Biotic and abiotic stresses increase NO (also an atmospheric pollutant) in plants. NO causes PTMs of proteins including *S*-nitrosylation and tyrosine nitration. *S*-Nitrosylation converts free cysteine to *S*-nitrosylated cysteine. This modification is commonly detected by biotin switch assay. Tyrosine nitration will nitrate tyrosine residues and is detected using antinitrotyrosine antibodies.

commonalities and differences in targets and extent of thiol modification in oxidative and nitrosative stress *in vivo*.

NO-mediated regulation of physiological processes and its contribution to important traits such as crop yield is gradually being worked out. Interesting and novel information about regulatory mechanisms such as NO-responsive proteins and their validation is getting accumulated at a rapid pace. Present observations suggest that NO could modulate carbon, nitrogen, and sulfur metabolism as well as important processes such as photosynthesis and glycolysis, emphasizing that it could be identified as an important regulatory mechanism such as phosphorylation. Studies on rice have already indicated that air pollution (smog) adversely affects crop yield [49]. Detailed analysis of NO-modulated proteins induced in smog-simulated conditions could provide us with a snapshot of protein modulations. This information could be useful in developing pollution-quenching crops in the future.

Acknowledgments

The research work was supported by a research grant to RD from The Council of Scientific and Industrial Research (38[1127/06/EMR-II), Government of India. JKA thanks CSIR for the research fellowship.

References

1 Mann, M. and Jensen, O.N. (2003) *Nat. Biotechnol.*, **21**, 255–261.
2 Huber, S.C. and Hardin, S.C. (2004) *Curr. Opin. Plant Biol.*, **7**, 318–322.
3 Lamattina, L., Garcia-Matta, C., Graziano, M., and Pagnussat, G. (2003) *Annu. Rev. Plant Biol.*, **54**, 109–136.
4 Neill, S.J., Desikan, R., and Hancock, J.T. (2003) *New Phytol.*, **159**, 11–35.
5 Grün, S., Lindermayr, C., Sell, S., and Durner, J. (2006) *J. Exp. Bot.*, **57**, 507–516.
6 Crawford, N.M., Galli, M., Tischner, R., Heimer, Y.M., Okamoto, M., and Mack, A. (2006) *Trends Plant Sci.*, **11**, 526–527.
7 Zemojtel, T., Fro7pf2pt´hlich, A., Palmieri, M.C., Kolanczyk, M., Mikula, I., Wyrwicz, L.S., Wanker, E.E., Mundlos, S., Vingron, M., Martasek, P., and Durner, J. (2006) *Trends Plant Sci.*, **11**, 524–525.
8 Thomas, D.D., Liu, X., Kantrow, S.P., and Lancaster, J.R. (2001) *Proc. Natl. Acad. Sci. USA*, **98**, 355–360.
9 Wendehenne, D., Pugin, A., Klessig, D.F., and Durner, J. (2001) *Trends Plant Sci.*, **6**, 177–183.
10 Sawa, T. and Ohshima, H. (2006) *Nitric Oxide*, **14**, 91–100.
11 Kalyanaraman, B. (2004) *Proc. Natl. Acad. Sci. USA*, **101**, 11527–11528.
12 Gow, A.J., Farkouh, C.R., Munson, D.A., Posencheg, M.A., and Ischiropoulos, H. (2004) *Am. J. Physiol. Lung Cell. Mol. Physiol.*, **287**, L262–L268.
13 Abat, J.K., Saigal, P.T., and Deswal, R. (2008) *Physiol. Mol. Biol. Plants*, **14**, 119–130.
14 Hess, D.T., Matsumoto, A., Kim, S.O., Marshall, H.E., and Stamler, J.S. (2005) *Nat. Rev. Mol. Cell Biol.*, **6**, 150–166.
15 Stamler, J.S., Lamas, S., and Fang, F.C. (2001) *Cell*, **106**, 675–683.
16 Taldone, F.S., Tummala, M., Goldstein, E.J., Ryzhov, V., Ravi, K., and Black, S.M. (2005) *Nitric Oxide*, **13**, 176–187.
17 Greco, T.M., Hodara, R., Parastatidis, I., Heijnen, H.F.G., Dennehy, M.K., Liebler, D.C., and Ischiropoulos, H. (2006) *Proc. Natl. Acad. Sci. USA*, **103**, 7420–7425.
18 Lindermayr, C., Saalbach, G., and Durner, J. (2005) *Plant Physiol.*, **137**, 921–930.
19 Romero-Puertas, M.C., Campostrini, N., Mattè, A., Righetti, P.G., Perazzolli, M., Zolla, L., Roepstorff, P., and Delledonne, M. (2008) *Proteomics*, **8**, 1459–1469.
20 Abat, J.K., Mattoo, A.K., and Deswal, R. (2008) *FEBS J.*, **275**, 2862–2872.
21 Wang, K.L.C., Li, H., and Ecker, J.R. (2002) *Plant Cell*, **14**, S131–S151.
22 Leshmen, Y.Y. and Pinchasov, Y. (2000) *J. Exp. Bot.*, **51**, 1471–1473.
23 Leshmen, Y.Y. (2001) *Nitric Oxide in Plants*, Kluwer Academic Publisher, London, UK.
24 Lindermayr, C., Saalbach, G., Bahnweg, G., and Durner, J. (2006) *J. Biol. Chem.*, **281**, 4285–4291.
25 Bright, J., Desikan, R., Hancock, J.T., Weir, I.S., and Neill, S.J. (2006) *Plant J.*, **45**, 113–122.
26 Lum, H.K., Lee, C.H., Butt, Y., and Lo, S. (2005) *Nitric Oxide*, **12**, 220–230.
27 Genkov, T., Du, Y.C., and Spreitzer, R.J. (2006) *Arch. Biochem. Biophys.*, **451**, 167–174.
28 Harrison, E.P., Willingham, N.M., Lloyd, J.C., and Raines, C.A. (1998) *Planta*, **204**, 27–36.
29 Almeida, A., Moncada, S., Bolanos, J.P., Cidad, P., and Almeida, J. (2004) *Nat. Cell Biol.*, **6**, 45–51.
30 Gould, K.S., Lamotte, O., Klinguer, A., Pungin, A., and Wendehenne, D. (2003) *Plant Cell Environ.*, **26**, 1851–1862.
31 Belenghi, B., Romero-Puertas, M.C., Vercammen, D., Brackenier, A., Inzé, D., Delledonne, M., and Breusegem, F.V. (2007) *J. Biol. Chem.*, **282**, 1352–1358.
32 Feechan, A., Kwon, E., Yun, B.W., Wang, Y., Pallas, J.A., and Loake, G.J. (2005) *Proc. Natl. Acad. Sci. USA*, **102**, 8054–8059.
33 Zhang, Y., Wang, L., Liu, Y., Zhang, Q., Wei, Q., and Zhang, W. (2006) *Planta*, **224**, 545–555.

34 Valderrama, R., Corpas, F.J., Fernandez-Ocaña, A., Chaki, M., Luque, F., Gómez-Rodríguez, M.V., Colmenero-Varea, P., Río, L.A., and Barroso, J.B. (2007) *FEBS Lett.*, **581**, 453–461.

35 Garcia-Mata, C. and Lamattina, L. (2001) *Plant Physiol.*, **126**, 1196–1204.

36 Garcia-Mata, C., Gay, R., Sokolovski, S., Hills, A., Lamattina, L., and Blatt, M.R. (2003) *Proc. Natl. Acad. Sci. USA*, **100**, 11116–11121.

37 Sokolovski, S. and Blatt, M.R. (2004) *Plant Physiol.*, **136**, 4275–4284.

38 Lee, U., Wei, C., Fernandez, B.O., Feelisch, M., and Vierling, E. (2008) *Plant Cell*, **20**, 786–802.

39 Jaffrey, S.R. and Snyder, S.H. (2001) *Sci. STKE*, PL1 (doi:10.1126/stke.2001.86.pl1).

40 Radi, R. (2004) *Proc. Natl. Acad. Sci. USA*, **101**, 4003–4008.

41 Wiseman, H. and Halliwell, B. (1996) *Biochem. J.*, **313**, 17–29.

42 Ischiropoulos, H. (2003) *Biochem. Biophys. Res. Commun.*, **35**, 776–783.

43 Aulak, K.S., Miyagi, M., Yan, L., West, K.A., Massillon, D., Cradd, J.W., and Stuehr, D.J. (2001) *Proc. Natl. Acad. Sci. USA*, **98**, 12056–12061.

44 Saito, S., Yamamoto-Katou, A., Yoshioka, H., Doke, N., and Kawakita, K. (2006) *Plant Cell Physiol.*, **47**, 689–697.

45 Bruckdorfer, R. (2005) *Mol. Asp. Med.*, **26**, 3–31.

46 Ludidi, N. and Gehring, C. (2003) *J. Biol. Chem.*, **278**, 6490–6494.

47 Netto, L.E.S., Oliveira, M.A., Monteiro, G., Demasi, A.P.D., Cussiol, J.R.R., Discola, K.F., Demasi, M., Silva, G.M., Alves, S.V., Faria, V.G., and Horta, B.B. (2007) *Comp. Biochem. Physiol. C*, **146**, 180–193.

48 Leichert, L.I., Gehrke, F., Gudiseva, H.V., Blackwell, T., Ilbert, M., Walker, A.K., Strahler, J.R., Andrews, P.C., and Jakob, U. (2008) *Proc. Natl. Acad. Sci. USA*, **105**, 8197–8202.

49 Auffhammer, M., Ramanathan, V., and Vincent, J.R. (2006) *Proc. Natl. Acad. Sci. USA*, **103**, 19668–19672.

Index

Nitric Oxide in Plant Physiology. Edited by S. Hayat, M. Mori, J. Pichtel, and A. Ahmad

ISBN: 978-3-527-32519-1

r

s